Watermills of Sussex: Vol 2 has been published as a Limited Edition of which this is

Number **184**

A list of original subscribers is printed at the back of this book.

FRONT COVER: The refurbished Ifield Mill in 1980 (TH)

Rowfant Mill in the early 1900's (NC)

WATERMILLS of SUSSEX

(VOLUME II - WEST SUSSEX)

BY

DEREK STIDDER & COLIN SMITH

PUBLISHED BY DEREK STIDDER AND COLIN SMITH
PRODUCED AND PRINTED BY PHEASANT COMMUNICATIONS LIMITED
LAYOUT AND REPROGRAPHICS BY TRIFFIK TECHNOLOGY LIMITED

Central © Derek Stidder & Colin Smith 2001

ISBN 0 9540071 0 7

The moral right of the authors has been asserted. No part of this publication may be reproduced, stored in a retrieval system or transmitted, in any form or by any means, electronic, mechanical, photocopying, recording or otherwise, without the prior permission of the Publishers.

Any copy of this book issued by the Publishers as clothbound or as a paperback is sold subject to the condition that it shall not by way of trade or otherwise, be lent, re-sold, hired out or otherwise circulated without the Publisher's prior consent, in any form of binding or cover other than that in which it is published, and without a similar condition including this condition being imposed on a subsequent purchaser.

Contents

1. ACKNOWLEDGMENTS 6
2. TRIBUTE TO FRANK GREGORY 7
3. THE SYDNEY SIMMONS RESEARCH NOTES 7
4. INTRODUCTION 9

THE WATERMILL IN WEST SUSSEX 11
THE RIVERS 12
WATERWHEELS AND WATERPOWER 15
TURBINES 16
TIDE MILLS 16
LAYOUT OF WATERCOURSES AND MILLPONDS 16
OTHER WATER POWERED INDUSTRIES 17
    PAPER 17
    LEATHER DRESSING 17
    FULLING 17
    IRON 17
A TYPICAL MILL - LURGASHALL MILL 18

5. THE MILLS
    RIVER ARUN 21
    RIVER ADUR 47
    RIVER ROTHER 69
    RIVER OUSE 99
    CHICHESTER AREA 113
    RIVER MOLE / RIVER MEDWAY 139
    RIVER WEY 147

6. CONCLUSIONS 151
7. GLOSSARY OF MILLING TERMS 152
8. SPAB 153
9. BIBLIOGRAPHY 154
10. INDEX 155
11. LIST OF SUBSCRIBERS 159

ACKNOWLEDGEMENTS

We are again indebted to the mill organisations and individuals that have kindly supplied us with information from both public and private sources. It particular, we must thank Mr R Hawksley for his constructive criticism and for providing additional research material, David Tomlinson for once again providing the maps and diagrams and to Priscilla Nobbs for her encouragement. Ted Henbery has kindly supplied information on the history and the restoration work on Ifield Mill, which is a prime example of what can be achieved by a dedicated group of volunteers.

The staff and members of many organisations have been of great assistance to us and they include, the Wealden Postcard Club, Sussex Industrial Archaeology Society, the Mills Section of the SPAB, West Sussex Record Office, Brighton Reference Library, and finally, the Science Museum Library for permission to consult the Simmons papers for Sussex and the Denis Sanders Collection. A word must also go to F. W. Mays & Co who have again unwittingly provided transport to all the 132 watermill sites mentioned in this book.

Also, grateful thanks are given to the mill owners who, with one notable and unfortunate exception, allowed us to roam freely about their property to record, measure and photograph.

The majority of these sites are on private property and permission to view must be obtained from the owners in advance.

As always, photographs play an important part in this book and we are very grateful, in particular, to Alan Stoyel, Roger Packham, Mrs Vivien Uwins, Douglas White and Ted Henbery.

Key to Caption Credits

AB	Alan Barwick		FG	Frank Gregory
AS	Alan Stoyel		NC	Nick Catford
BLSL	Brighton Local Studies Library		RP	Roger Packham
DW	Douglas White		PA	Peter Allen
EH	Ted Henbery		VU	Vivien Uwins
MB	Margaret Baldock			

All other photographs are from the authors' collection.

FRANK WILLIAM GREGORY
3rd November 1917 – 7th June 1998

The passing of Frank Gregory has left a great void in the world of windmills and watermills. Frank had an extensive knowledge, not just of his beloved Sussex, but of the whole country and abroad, and as he said once, 'there was never too much that you could know about mills'. Frank was always willing to help fellow mill enthusiasts with his extensive knowledge gained from an early age. On the many times that Colin and I met in his front room at home (those who had the pleasure of visiting him will have a chuckle about this) seemingly 'buried' beneath his endless piles of mill related books and papers, he was always willing to find that photograph or remember that reference. Once I sent him a rather obscure photograph of a 'Sussex Watermill', which I could not identify, and back by return of post came the name of the mill (in this case Rowfant) along with a copy of one of his famous ink sketches. Frank was always like that with most of the information being in his head, for he rarely wrote anything down.

Although he lived a busy lifestyle and travelled widely, he took time to proof read our manuscript 'The Watermills of East Sussex' and took more time to add or correct details to it. He had offered to proof read this book but, alas, time beat him. One of my fondest memories of Frank was when I met him at an open day at the restored Cobham Mill in Surrey on a Sunday afternoon, or rather passed him as he was hurrying off to visit yet another mill. But even then he took time for a chat and asked me how the East and West Sussex books were progressing and again offering to help – but that was Frank, a true gentleman and supreme mill enthusiast, who is sadly missed.

HERBERT EDWARD SYDNEY SIMMONS
28th September 1901 – 26th October 1973

Sydney Simmons, as he was more affectionately known, was born at Washington in West Sussex, and had a lifelong interest in both windmills and watermills. Over the years he visited countless mill sites and consulted and recorded a wide range of documentary sources. He also collated further information on mills that he was unable to visit. By the end of his life, he had produced an extensive collection of manuscripts and research notes arranged by county, together with annotated maps and over 2,000 photographs.

In January 1974, his widow, Mrs E. Simmons bequeathed his collection to the Science Museum Library in London where his notes were transcribed and typed into an easy and accessible format, arranged individually by county. The museum's photographic department also made copies from his original negatives, which are now housed in the National Monument Record Centre at Swindon.

The Simmons Collection may be consulted in the Library by anyone holding a current reader's ticket. The research notes are extensive and provide a reference document for both mill enthusiast and those engaged in historical research. We are indebted to his documentary reference work during our investigation into the history of the watermills in Sussex.

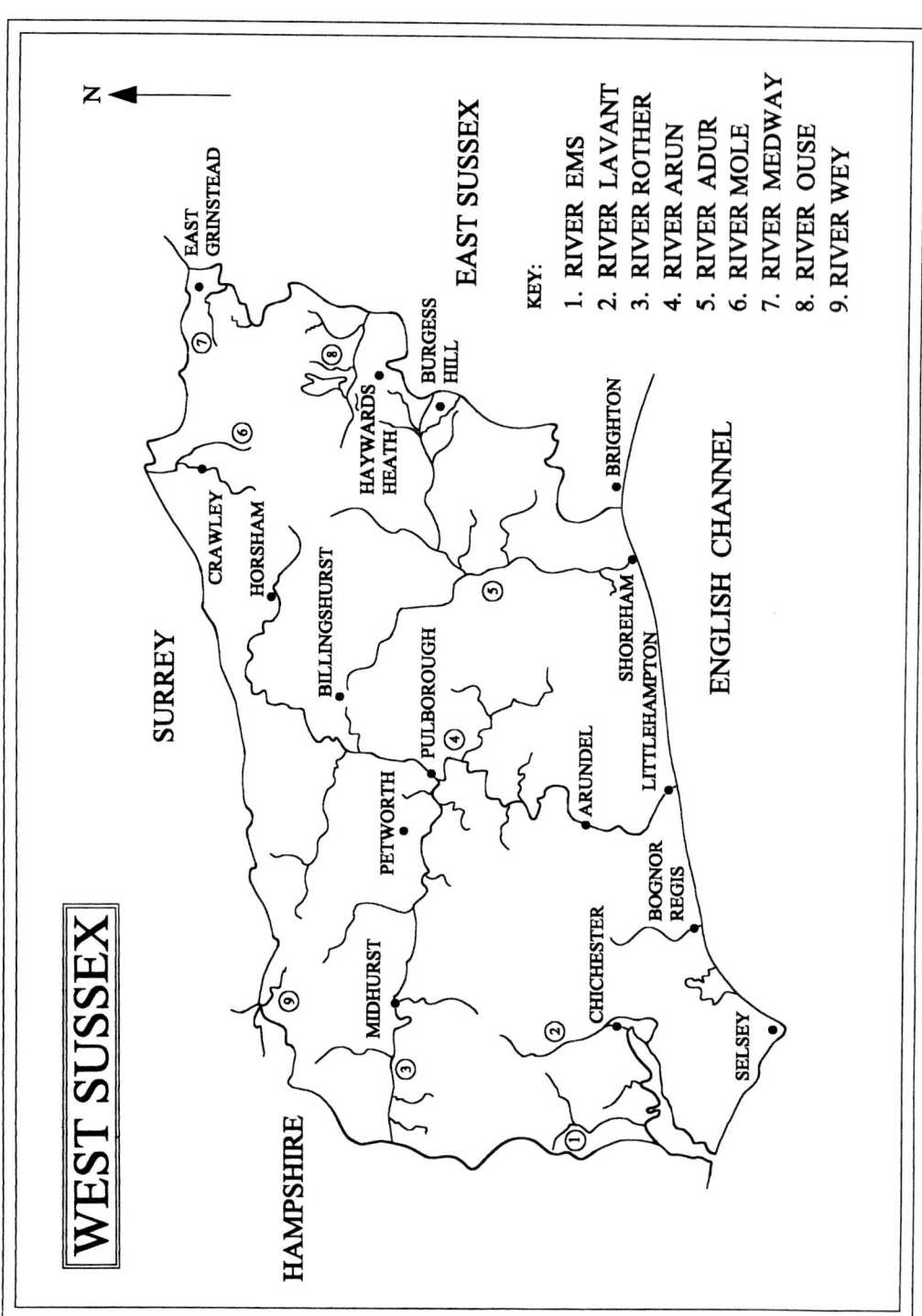

INTRODUCTION

The county of Sussex has been divided into two administrative areas for many years but the characteristics of each are far apart, with the deep wooded valleys of East Sussex not found in West Sussex, where rolling downland is more common. The county towns of Lewes and Chichester vie for importance but many residents rue the division of Sussex for administrative purposes. Furthermore, arising from the 1972 Local Government Act, former important East Sussex towns such as Haywards Heath, East Grinstead and Burgess Hill were transferred into West Sussex, Crawley having been transferred previously.

The coastline of West Sussex stretches from west of Portslade by Sea in the east, to Hermitage in the west, while inland, its boundary is shared with Surrey and Hampshire. Although the South Downs dominate the landscape, Sussex is one of the most wooded counties in the country, and together with its open tracts of land, and seaside towns such as Worthing and Bognor Regis, it is a county of contrasts. Inland, behind the South Downs, are the fields and lanes of the Sussex Weald with picturesque villages grouped around village greens. The geology of the county is dominated by the chalk of the South Downs, under which there are layers of Upper and Lower Greensand, with the 'elasticity' of the Weald Clay, found in the county, 'encouraging' the use of the navigable waterways for the transportation of goods. The south-west of the county has a variety of sheltered waterways, harbours and quiet creeks, which formed an ideal environment for the deployment of the seven tide mills that formerly worked here. The principle employment in the county was agricultural based and even as late as 1947, it accounted for 23% of the population in the Horsham area for example, while countywide, this figure has now been reduced to 5%.

There are eight rivers of some substance in the county and apart from the River Rother, which connects Petersfield with Pulborough, all others flow southwards to the sea. The arrival of the railway was of great benefit to the various mills, as it changed the social attitudes and habits of the population, and provided for an ease of transporting flour throughout the area. Long past the days of milling soke, the millers and owners prospered, as the population of the towns and the villages expanded, with the railway network enabling quicker and more convenient carriage of goods. This encouragement to mill owners in the middle of the century was short lived due to the repeal of the Corn Laws in 1846. This allowed the import of foreign wheat from every corner of the world and, to cope with this, large steam powered roller mills were set up, many within the Port of London itself. Wheat arriving from abroad was milled and dispatched in such volumes that the rural watermill could in no way compete commercially.

Another reason for the decline was the introduction of the roller milling system which could produce vast quantities of flour quickly, when compared to the output from the traditional stone ground mill. This system had been developed on the Continent and by 1880, its advantages had became apparent in this country as it could produce white flour and was ideal for grinding the harder imported wheat. In an attempt to survive, several mills in West Sussex installed the new roller system, such as at Emsworth Town Mill, but the majority were not converted, and had to cease flour milling and just solely rely on producing animal feed. In the latter years of the 19th century, watermills closed down in their thousands across the country and West Sussex did not escape this trend as mills such as Kirdford Mill succumbed to the inevitable. There were exceptions to this with Storrington Town Mill continuing grinding using traditional methods and water power until its closure in the 1950's.

As with the book 'The Watermills of Sussex - Volume 1 East Sussex, a similar format has been adopted in that general watermill history and methods of operation, have been kept to a minimum. Also, as with East Sussex, the principal water powered industry was flour milling although there is now no working commercial mill. The resited Lurgashall Mill, at the Weald and Downland Museum, although milling a certain amount of flour, does enable visitors to see a working watermill in action. To a lesser extent, paper making was a principal industry, but nothing remains with even Shottermill New Mill converted to flour milling when the paper mill industry closed. The Sussex Iron industry affected sites such as Shillinglee Mill, but was not extensive in comparison to East Sussex.

We have managed to visit and locate 132 water powered sites in the county but 83 are no more than sites with few remains while, of the remaining 49 mills, many have been converted into houses, such as Cocking Mill, while others became offices such as Warnham Mill near Horsham. Only Ifield Mill and Horsted Keynes Mill are capable of working (not commercially) although Dean's Mill at Lindfield and Cobb's Mill could both easily work again if their waterwheels were repaired.

It was not the purpose of this book to compile every known fact about the history of a mill site. We have concentrated on a period from about 1700, when documentary records (such as Fire Insurance Policies) begin to exist, which specifically relate to individual mill owners and tenants. Where earlier history exists, it has been included as felt appropriate to the individual mill report. Hopefully, this will encourage people to make further in depth research into what was once one of the most important buildings in any community.

This book chronicles, for the first time, the history of the watermill industry of West Sussex, the people who worked them, and their development and decline over the centuries.

Delivery lorries at Midhurst North Mill in 1958 (MB)

THE WATERMILL IN WEST SUSSEX

The majority of watermills in the county were small and normally contained 2 or 3-pairs of stones. Naturally, there were exceptions to this, notably Horsham Town Mill and Brewhurst Mill, both having flood wheels which could operate during high river levels. All the mills were built in local materials and many were timber framed with brick to the first floor, a fine example of this being Horsted Keynes Mill. However, with the introduction of the brick tax, new mills reverted back to a more conventional timber framed construction but Emsworth Town Mill, rebuilt after the abolition of the brick tax, is a prime example of the late Victorian style of architecture. Up to the beginning of the 19th century, elm and oak were the principal hardwoods used in mill construction, with the former, with its water resistant properties, predominately used for waterwheel construction. The advent of cast iron revolutionised the construction of machinery and most mills converted to this at some time. The introduction of the roller mill system required the appropriate structural strengthening of a mill and usually only brick built mills could cope with this development, but there is little evidence of this in the county. Of the 132 mills identified in West Sussex not one is working commercially, and it is only at the Weald and Downland Museum that a working mill can be seen operating. Mills were often sturdily built and situated in picturesque surroundings, sometimes with millponds, and many were accordingly converted into houses, the machinery rarely retained. Other mills were just demolished for their building materials and many sites that had been utilised continuously for hundreds of years came to an inglorious end in the nineteenth and twentieth centuries.

Cobb's Mill in working order in 1953

THE RIVERS OF WEST SUSSEX

The county of West Sussex was dependent upon the use of its navigable waterways for the transport of buildings materials and general goods, owing to the poor state of the highways. These waterways meander through green meadows and wide valleys before piercing the South Downs to the English Channel, with the River Rother the only exception.

As with East Sussex rivers, the streams and tributaries within West Sussex were used extensively to drive the majority of the 132 recorded watermills. The three principal rivers were the Arun, Adur and the Rother, with the latter having the most mills in total. All were navigable to a certain extent, but were improved at the beginning of the 19th century during what was recognised as the great canal building age. The first 'canals' in West Sussex were, strictly speaking, navigations that involved the removal of sandbars and the straightening of the meandering course of the rivers. The true canal was a new waterway, such as that linking the navigation at Pulborough through to the Surrey/Sussex border and to the Wey navigation with 14 locks. The waterways were at their busiest in the 1830's with 36,000 tons of cargo carried on the Wey and Arun canal, while in 1839, with the last canal constructed between Ford and Portsmouth was built between 1817-23. Apart for providing a direct route from London to the Naval dockyard at Portsmouth and beyond, the canal system enabled millers to buy and sell grain and flour over a much wider area. The decline of the canals and the navigations, and the smaller watermills, was due to the opening of the railway which spread its tentacles throughout Sussex, the freight being transferred to the 'rail road' as it was then known.

THE RIVER ARUN

This is one of the principal waterways of West Sussex that shares its source with Surrey. For much of its length, the river ran parallel with the Wey and Arun Navigation especially so, above Pallingham Quay (TQ 036 214).

The river rises at two separate locations, Chiddingfold in Surrey and Mannings Heath near Horsham, with its confluence south east of Loxwood. There were 10 mills above Billingshurst of which 6 were grouped around Horsham. After the convergence of the two feeder streams, the river continues to Pulborough where it joins the River Rother and from this point, is tidal, passing through Bury, Houghton, Arundel before entering Littlehampton Harbour.

In many respects this was a good watermill river, with a constant head of water, but below Pulborough, its tributaries had to be used instead. There were 27 mills situated on, or adjacent, to the main river with most of the larger mills on the upper reaches, such as the Town Mill at Horsham. The River Arun was made navigable to Pulborough during the reign of Elizabeth I and by 1637, barges could easily reach Pallingham Quay, toll free, after which the Navigation was extended to Newbridge, west of Billingshurst, in 1787 passing through 4 locks. Later in 1816, it was extended through to the Wey Navigation at Shalford. The upper part of the Navigation was first affected by railway competition in 1857, when the Mid-Sussex Railway was authorised to build a line from Horsham to Pulborough and Petworth, and later to Midhurst, opening in 1866. Although the waterway competed for many years, the closure order was made in 1896 with barge traffic continuing on the lower reaches on a limited scale until the 1920's.

THE RIVER ADUR

This river was carrying traffic on a very limited scale in the 18th century, but with the success of the nearby Arun Navigation, an Act was made in 1807 which enabled barges drawing a depth 3ft or less, to work on the tidal River Adur between Shoreham and Bines Bridge. Later the Baybridge Canal, following the river course, extended the navigation through to the Worthing-Horsham road (A24) at West Grinstead, with 3 locks.

The principal stream of the western branch flows through Shipley and West Grinstead. The eastern branch has several streams feeding into it, one of which rises at Ditchling Common with the two branches meeting north west of Henfield. There was a profusion of mills on the eastern branch (24 in total) and on a very small tributary, south of Burgess Hill, were three large watermills ie.Cobb's Mill, Hammonds Mill (demolished in 1975) and Ruckford Mill. There were only 3 mills on the western branch, all of which were small.

RIVER ROTHER

This river is one of the shortest in West Sussex, rising on the slopes of Noar Hill near Selbourne in Hampshire, but there were 30 mills working throughout its length. Most of the mills were small or average size with Midhurst North Mill one of the largest extant buildings. The river enters the county east of Petersfield and flows in an easterly direction passing Midhurst to the north and Petworth to the south to its confluence with the River Arun at Stopham.

The western River Rother was navigable to Fittleworth in 1560, but in 1781 Lord Egremont of Petworth House, obtained an Act to make the river navigable to Midhurst (1794), with a branch to Petworth (1795). The line surveyed by William Jessup followed the natural course of the river, rising 86 feet from Stopham to Midhurst, passing through a series of 8 locks. The short branch northwards to Petworth was built partly on an embankment with 2 locks, but was never a success. The main navigation was a commercial success with coal and chalk imported, with flour and timber going the other way. The navigation was of great benefit to the citizens of both Petworth and Midhurst and to millers and farmers alike. It became cheaper to use the navigation to Stopham Bridge than pay for the transport previously employed. However, the railway to Midhurst opened in 1866 and took most of the trade away, although the navigation survived, working commercially until March 1888.

RIVER OUSE

The Ouse is the longest river in Sussex but only a small length is within West Sussex. The source of the river is on the edge of St.Leonards Forest with most of the 13 mills situated on its tributaries, with Horsted Keynes Mill a fine working example of its type. The most well known mill is Dean's Mill at Lindfield, which is complete and with some attention to the buckets of the waterwheel, could easily work again (it stopped in 1976).

The Ouse Navigation was extended up to Upper Ryelands Bridge, north of Haywards Heath in 1812 with 8 locks within West Sussex and was very successful until the arrival of the railway.

CHICHESTER AREA

In a five mile radius from Chichester, there were 24 mills on a varying assortment of small rivers, tributaries and streams. Many of these mills have gone with the large tide mill at Sidlesham no longer dominating the flat landscape on the edge of what is now the Pagham Harbour Nature Reserve. Some of the largest mills in the county were found in this region, such as West Ashling Mill and Ratham Mill. The majority of rivers flowed into Chichester Harbour with the tributaries serving Runcton Mill, Oving Mill and Aldingbourne Mill the exceptions. One of the most interesting is Bosham Mill with its two overshot waterwheels.

On the western extremity of the county is the River Ems, a small river that rises near the village of Stoughton, and flows southwards into the western channel of Chichester Harbour. Of the 5 mills within West Sussex (Quay Mill being in Hampshire), Emsworth Town Mill, Lumley Mill and Westbourne Mill were large and powerful mills requiring millponds of varying sizes, while the Old and New Slipper Mills were tide mills.

RIVER MOLE

This is predominately a Surrey river, crossing the border at Gatwick Airport, from the Crawley area. Of the 5 mills, east and south east of Crawley, Hazelwick Mill succumbed to the requirements of the New Town development, while Rowfant Mill is now a house. On the opposite side of Crawley, is Ifield Mill that was derelict in the early 1970's and devoid of machinery. Over the following 26 years a loyal band of volunteers have painstakingly restored the mill, often using machinery saved from demolished mills in the county, (predominately Hammonds Mill demolished in 1975). A large millpond, bisected by the Crawley-Horsham railway, provides an ample water supply. The site of nearby Bewbush Mill has almost been lost to vegetation.

RIVER MEDWAY

As with the Wey, Ouse and Mole, only a small section of the River Medway, with 3 mills, was within West Sussex. There is no trace of Warren Furnace Mill and both Dunnings Mill and Fen Place Mill have been converted to other uses. At all three location extensive ponding arrangements had to be made.

RIVER WEY

The River Wey is shared between Hampshire and Surrey apart from a small tributary rising in Sussex, south east of Haslemere. Three of the five mills in Haslemere had association with West Sussex, often with the county boundaries mered to a section of the waterways at each site, apart from Shottermill that was located within West Sussex. Millponds had to been constructed to establish a constant head of water from the small tributaries.

Waterwheels And Water Power

WATERWHEELS

The waterwheel was the power source for every watermill, until the advent of steam mechanisation during the late 19th century. Other forms of power included diesel engines, turbines, gas and later, electricity. The 50hp Tangye gas engine at Cobb's Mill is a rare survivor of a time when the river levels started to fall and auxiliary power was necessary. At the time of the Domesday Survey, the geared watermill was in existence and its methodology remained virtually unchanged thereafter. Initially, the mills would have been small with wooden machinery worked by a wooden undershot waterwheel driving 1-pair of millstones, but as millers searched for more efficiency, so other methods of water supply and increased use of power were introduced. It was found that if the water was passed over the top of a waterwheel it proportionally increased its efficiency and certainly by the 16th century, millponds were being constructed to power this type of wheel. Warnham Mill is one of the largest millponds in West Sussex owing its origins to the iron industry, but its subsequent use for a corn mill was a common development. While this method of ponding required civil engineering works, other mills such as Wanford Mill, on the River Arun, relied on a simple but direct supply from the river.

As with East Sussex, the majority of waterwheels were externally mounted, although the two extant overshot waterwheels at Ruckford Mill were later housed in a brick extension as a protection against the weather. The arrival of cast iron provided the miller with the opportunity to change over to a material that was more durable than wood and, subsequently, many waterwheels were constructed in this relatively new type of material. Within West Sussex there were three main types of waterwheel, classified by the entry and position of the water onto the wheel.

Overshot

The overshot waterwheel was the most common type of wheel used throughout the county. This type of wheel requires a greater fall of water than its diameter and in most cases, requires extensive engineering works and a millpond. Water is channeled to a point just over the top axis of the wheel into closed sided buckets and it is the weight of this water which keeps the wheel in motion. The 18ft x 3ft 4in diameter waterwheel at Horsted Keynes Mill is a superb working example of its type, but for sheer power the massive cast iron overshot waterwheel at Ifield Mill takes some beating when turning. Throughout the county, large millponds were required to keep the mills working on average between 6 - 8 hours a day.

Breastshot

For this type of wheel, the water is regulated onto a point level with the axle of the waterwheel and more often than not, required a millpond. Either buckets or curved vanes were used but with the latter, the most common type, there was a certain amount of water loss. The two external breastshot waterwheels at Rowner Mill (sadly demolished) were fine examples of its type.

Undershot

This is the least efficient but simplest type of waterwheel, which is just placed in a channel and worked by water passing just above the vanes at the bottom of the wheel. It was commonly found in low lying areas where there was a constant head of water, and although little engineering work was required it was rarely used in the county. This type of wheel was used at Lodsbridge Mill and Hardham Mill, on the River Rother, where a short leat powered two undershot waterwheels working together in parallel.

TURBINES

Turbines were introduced into the country during the 19th century, but surprisingly, did not make a great impact into West Sussex. One of its disadvantages was that on occasions, its installation required alteration to the mill and to its method of transmitting power to the drive machinery. Undoubtedly it used less water than a conventional waterwheel but nevertheless, only a few examples of a turbine can be traced. Ratham Mill installed a turbine to drive a generator to power the electrically driven machinery in the mill, but even here, the waterwheel was retained for hoisting purposes and for working an oat crusher. The lack of use of turbines in the county is puzzling, when compared to Hampshire, but it was the reluctance to change (and its expense!) that may have been the fundamental reason at a time when millers were suffering from the economic effects of roller mills and steam mills.

TIDE MILLS

Tide mills are usually built on inlets branching off tidal estuaries and incorporated a 'sea hatch' which was a special type of sluice gate. The sluice gate opens with the tide and automatically closes at high tide after which the impounded water is released and milling continues until the tide comes in again, averaging, about 4 - 5 hours of milling from each tide.
In the area around Chichester was found one of the main concentrations of tide mills in the country as, during the 19th century, there were seven tide mills working, six on the edge of Chichester harbour and one at Pagham harbour. The only mill building to survive is at Birdham, which closed in 1935, and was a mill where vessels of 150 tons could deliver corn directly to the mill. Of the other mills, there are few remains, apart from some dried up millponds. The mill at Sidlesham was exceptionally large and was steam assisted in its latter working years and could be seen from miles around due to the flat landscape, until it was demolished in 1919. Of New Slipper Mill at Hermitage, there is no trace while at the site of Old Slipper Mill, at Old Fishbourne, the brick and slate storehouse has been converted into dwellings. The mills at Nutbourne and the Old Salt Mill at Bosham have disappeared without trace.

LAYOUT OF WATERCOURSES AND MILLPONDS

The site of each watermill was certainly not selected by chance as a range of considerations such as demand, water supply and water rights had to be investigated before expense was incurred. The downland landscape of West Sussex led itself to ponding arrangements at most mills to retain the slow moving rivers and streams. Of the 132 recorded mills, at least 85 were driven by ponded water with the overshot waterwheel in predominant use. At many sites the mills were often built into an embankment such as Burton Mill, while of the remaining sites many were driven directly from the river such at Gibbons Mill on the River Arun, a prime example.

OTHER WATER POWERED INDUSTRIES

Unlike East Sussex, the county had little association with water powered industries other corn milling. While its sister county had sites producing gunpowder, iron, paper, wool fulling and linseed oil, West Sussex had only minor dealings with paper, leather dressing, iron production and fulling.

Paper

In comparison to Surrey and to Kent in particular, the paper making industry was confined to just six sites, which is strange bearing in mind the purity of the water in the western part of the county. Dean's Mill near Lindfield, was producing paper in conjunction with Sharp's Paper Mill near Newick using the Ouse Navigation for transportation. There were two paper mills at Iping, while the other sites at West Ashling (1823 - 1850) and Duncton Paper Mill were producing either coarse or blotting paper. Of the five water powered sites at Haslemere only the New Mill had brief associations with the trade.

Leather Dressing

The regional centre for leather dressing in this area was in and around Godalming, while at the New Mill at Haslemere, Edward Masters was recorded as a leather dresser in 1882, using it in conjunction with the nearby Pitfold Mill (Surrey) until it closed down in 1902. This and Midhurst South Mill were the only sites in the county.

Fulling

There were fulling mills recorded at Duncton Dye Mill (after 1724) , Midhurst South Mill, Halfway Bridge Mill, Ardingly Mill and Cuckfield Upper Mill but little is known about their history.

Iron

There are sites in West Sussex, which had associations with the iron industry according to Ernest Straker's definitive book Wealden Iron. One was at Shillingllee Mill where a vast millpond testifies to its past use, similar to Warnham Mill.

A TYPICAL MILL - LURGASHALL MILL

Lurgashall Mill is a brick and timber framed three story building dating from the early years of the 18th century, although it was modified in later years with the tiled roof half hipped at one end to provide extra storage space. The source of water power was of importance to the commercial success of any watermill and especially here, as two overshot waterwheels worked together in tandem. To accommodate a plentiful supply of water, a large millpond was constructed with the mill set into the embankment to allow for the waterwheels to be used. The last remaining waterwheel came originally from Coster's Mill at West Lavington and is now working at its third location! There were once two separate sets of mainly old machinery working off each wheel, each driving 2-pairs of millstones. The mill has since been resited as a working exhibit at the Weald and Downland Museum at Singleton.

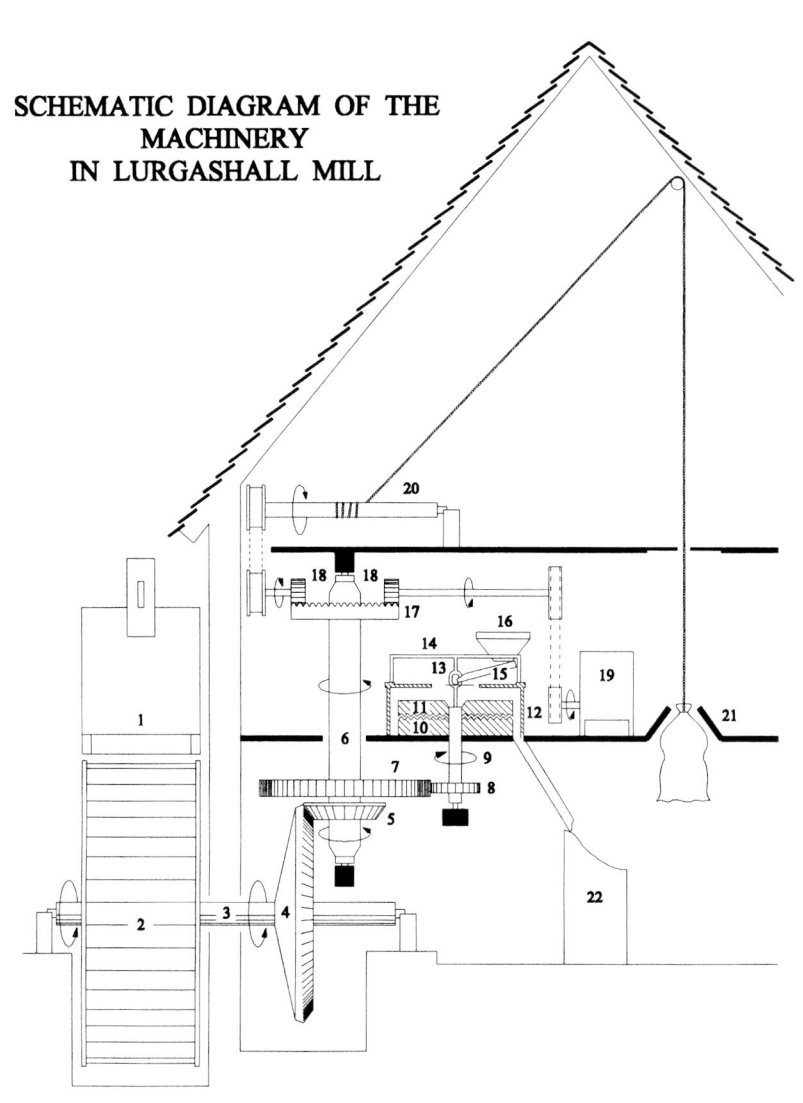

SCHEMATIC DIAGRAM OF THE MACHINERY IN LURGASHALL MILL

1. PENSTOCK
2. WATER WHEEL
3. WATERSHAFT
4. PIT WHEEL
5. WALLOWER
6. MAINSHAFT
7. GREAT SPUR WHEEL
8. STONE NUT
9. SPINDLE
10. BEDSTONE
11. RUNNER STONE
12. TUN
13. DAMSEL
14. HORSE
15. SHOE
16. GRAIN HOPPER
17. CROWN WHEEL
18. ANCILLARY DRIVES
19. FLOUR DRESSER
20. HOIST
21. SACK FLAPS
22. MEAL BIN

Above: The overshot waterwheel at Lurgashall Mill in 1949
Left: The floodwheel in 1957 at Brewhurst Mill
Below: Breastshot waterwheel at Dean's Mill in 1994

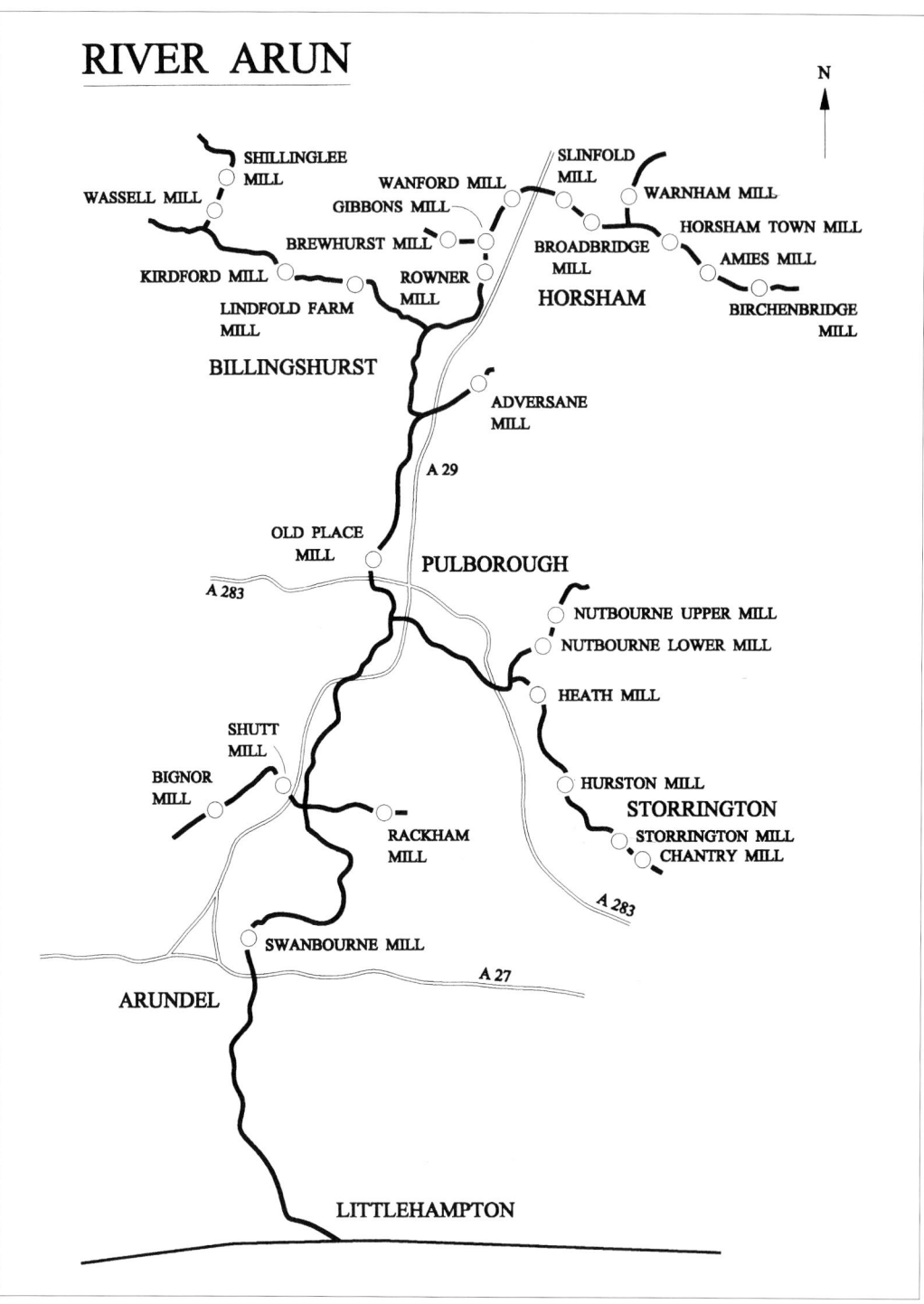

River Arun

ADVERSANE MILL *Adversane*
Tributary to River Arun - TQ 079 235 – East side of railway crossing.

The history of the site at Adversane is unknown and has disappeared into obscurity. Budgen marks his 1724 map as 'Oil Mill' but nothing else corroborates this. There is little to indicate the exact location apart from some old earthworks, through which a small stream meanders.

AMIES MILL *Horsham*
River Arun – TQ 183 291 – By Amiesmill Farm

This is an ancient site dating back to at least 1410 when it was listed as 'Assheles Mille', while a survey of 1650 refers to it as Amies Mill. The survey also stated that the mill, lately rebuilt, was in a good state of repair, with the mill house in decay.

There are no traces to be found to indicate the site and it is not known when the mill disappeared but it was marked on the 1813 Ordnance Survey 1" map but not mentioned in the 1852 Tithe Apportionment. However, a visit to this site is not without interest, as the small waterwheel of later farmers own-use mill remains. As with many other own-use mills, this was not situated on the site of the watermill but formed part of the nearby farm outbuildings.

The diameter of the waterwheel is small, difficult to determine and buried in a mass of tangled undergrowth. It was manufactured for Messrs H & E Lintoft in 1897 by 'Steele and Dobson' at their iron and brass foundry near Horsham railway station.
This own-use mill ceased working in 1928, and while the waterwheel is still in a reasonable condition, it will surely not last.

BIGNOR MILL *Bignor*
River Arun – TQ 982 142 – North of the village

This was not a large mill but unusually is situated on the side of a steep hill. The site is that referred to in the Domesday Survey, while another later reference concerns a dispute about the ownership of a messuage, watermill and land at 'Bygnor' Richard and then William Marshall were in occupation from at least 1845 until William's death in 1883. Joseph Butcher, T.Hedgecock, and finally, Edward Ede followed them until 1907, after which commercial milling ceased, although afterwards it was used occasionally by Bignor Farm.

In its latter working days, only 2-pairs of stones were used and it was working as a flour mill at the turn of the century when the ownership of the mill passed from the Slindon estate to the Bignor estate.

The mill is built to three floors of brick to an unattractive design, and a stone tablet 'AN 1844' records its date of construction. It was converted to residential accommodation in the 1930's when the machinery was removed and although the water supply to the mill has been diverted, the millpond, recently cleared out, forms a most attractive feature. A dummy wooden waterwheel has been built inside for aesthetic purposes.

BIRCHENBRIDGE MILL *Nuthurst*
River Arun – SU 193 291- South side of Horsham to Cowfold Road

The three buildings that made up Birchenbridge Mill, were of totally different building styles, that together were not the most attractive in Sussex.

The site is situated adjacent to the south side of the A281 Horsham to Cowfold road opposite a very sizeable millpond, constructed for the iron industry, with the road acting as a dam of such height, that the roof of the mill was barely above road level.

There was a corn mill in Sedgewick Manor in 1326, and this was replaced by an iron forge, working in 1598, when the Crown leased land to dig locally for iron ore, but the forge had gone by 1627 and was replaced by a corn mill.

The middle of the three buildings, the original mill, was built of brick and stone to three floors, with an iron overshot internal waterwheel and a penstock, manufactured by W.Cooper and dated 1865 that, in its latter days, powered a lathe. When this mill was added to on its eastern side, a second overshot waterwheel was fitted to work a sack hoist and ancillary machinery. Both waterwheels were about 14ft in diameter. The machinery was not of the standard arrangement and consisted of unusual drive equipment not commonly found in most watermills.

The sale particulars in the *Sussex Advertiser* of the 29 May 1824, gives an insight into the composition of the mill site. According to the report it had a drying kiln, two overshot waterwheels, 3 pairs of stones, was built to four floors, stables, a millwrights shop and a powerful head of water of over 12 acres. The mill, at the time, was in the occupation of Mary Tobutt, whose family had been at this mill since 1732, but following a dispute of the legality of the existing lease, she left soon after. An entry in the *London Gazette* of June 1778 lists John Tobutt as an insolvent debtor, and a prisoner at Horsham Gaol, but obviously the family carried on according to the 1824 sale notice. The mill and house were not sold at the auction but were later purchased for £3,400 by James Bravery who was the miller at Westcott Mill near Dorking. After a succession of millers the mill was later purchased by Charles Scrase-Dickens JP who leased it to James Killick with George Sharp in control. When the St.Leonards Forest estate was advertised for sale in 1878, Sharp was paying an annual rent of £15 per annum for the use of water obtained from the milllpond. According to *Kelly's* 1882 directory, Sharp was still here but soon after Charles Agate took over until he left in 1889 to work Ifield Mill. The mill was never re-let and Charles Mitchell managed the mill on behalf of Scrase-Dickins until milling ceased in the early 1900's, after which one of the waterwheels was converted to drive a dynamo for charging the batteries at Coolhurst, and for pumping water.

By 1946, the mill was disused and was demolished in 1958 with the old oak framed sluice gates removed in 1978. Water from the millpond still crashes down into the tumbling bay but apart from this, the disappearance of Birchenbridge Mill is complete.

BREWHURST MILL *Loxwood*
Tributary to River Arun - TQ 046 311 – Adjacent to Brewhurst Lane

The mill that stood here towards the end of the 19th century was a large three storey sandstone building with two waterwheels, but on the 23 August 1899 the upper floors of the mill were totally destroyed by fire according to a report that appeared in *The Miller*.

'On Wednesday August 23, the Water Corn Mill owned and occupied by Messrs Botting Bros. and known as Brewhurst Mill, Loxwood, Billingshurst, Sussex, was totally destroyed by fire at about 2.15pm. The mill, which for years past has been assisted by a steam engine, had recently been fitted with a petroleum oil engine, which was the cause of the fire. The mill was of brick and timber and contained 3 or 4 pairs of stones'

The mill was rebuilt utilising the ground floor brick base and was of four storeys high, with an outside lucomb for the first and second floor. Fortunately both waterwheels were undamaged by the fire and continued to power the new mill.

The earliest reference to a mill was in 1556 when John Myll was the occupier but, thereafter, little is known until the publication of directories in the 19th century. According to *Kelly's* 1851 directory, Henry Botting was the proprietor and it remained in the family until 1918, despite the fire in 1899. Thereafter, the mill was in the occupation of the 'Brewhurst Milling Co.' under the control of Walter Roper, and continued until 1938 (water power stopped here when the axle shaft of the overshot wheel broke in 1932). Earlier in 1928, a 41hp Blackstone Horizontal single cylinder diesel engine was installed and drove the mill via a fast and loose pulley system until commercial milling ceased in 1968. Occasionally, corn was milled for 'domestic' use until 1981.

Although the mill is disused there are still in existence the three sources of power that once drove the mill. The internal iron overshot waterwheel 10ft in diameter by 10ft wide set on a steel shaft under the ground floor of the mill. The waterwheel has suffered over the years and only the three sets of arms remain. The 8ft 8in diameter pit wheel and the cast iron upright shaft remain along with a large 9ft diameter spur wheel. Outside is an iron breastshot flood waterwheel, 14ft in diameter by 4ft wide, set on a wooden axle shaft and virtually complete, with the inscription 'King, Millwright, Rudgwick, 1861'. This flood wheel is connected to a 9ft diameter pit wheel projecting into the ground floor gearing to a horizontal shaft with a pulley which worked the existing 2-pairs of millstones, both 3ft 6in in diameter set in round wooden tuns. The second floor carries four large grain bins; belt driven oat crusher, winnowing machine; bucket type grain elevator and an extractor fan near to the main stone feed hoppers at the southern end, while the top floor consists of the service walkway between the main meal bins.

Ordnance Survey maps show that the mill was served by a large millpond with the tailrace flowing into a stream that runs into the Wey and Arun Canal which passes close by. The millpond is now completely overgrown and some recent plans to renovate this attractive mill appear to have fallen through.

BROADBRIDGE MILL *Broadbridge Heath*
Tributary to River Arun – TQ 144 304 – At end of track near Broadbridge Farm

This was a large watermill situated 2 miles west of Horsham, which had two waterwheels, which probably ceased working following problems with the water supply. This is an ancient mill site dating from 1298 while the first documented reference to it appears on the 1813 Ordnance Survey 1" map.

During most of the 19th century the mill was operated by the Stanford family and must have been a profitable mill with its 6-pairs of stones. It was built to four floors; the lower two of brick and the others of white weatherboarding and housed two waterwheels, north and south, with a turbine replacing the latter in 1882.

The north wheel was iron, 8ft 6in in diameter by 8ft wide, keyed to a small 6ft 6in pit wheel with the remaining machinery all iron powering 3-pairs of stones. A portable steam engine was used to supplement this waterwheel at times of water shortage. The south wheel, superseded by a turbine, drove the remaining pairs of stones. At this time, John Stanford advertised for sale a waterwheel 8ft in diameter by 5ft wide in an attempt to 'make room for more power', ie the turbine. Alfred Trim had taken over the mill by 1899, but a year later he was declared bankrupt and the Council took over the ownership of the mill in connection with their adjacent sewage works. Later in 1928, John Hole was working the mill and it was reported that he was doing a great deal of work for local farmers until closure, the machinery being removed in 1950.

The mill was demolished in the 1960's and only brick footings remain to the west of the track to the mill house where the old tailraces from the two waterwheels can be seen. The millpond, idyllic as ever, was used in the 18th century as a baptismal pond by Horsham Baptist Church.

CHANTRY MILL *Sullington*
Tributary to River Arun – TQ 092 138 – East side of Chantry Lane

Chantry Mill, (or Park Mill as it was sometimes called) together with the mill house, forms a most attractive group of buildings and although the former mill contains no machinery, the millpond and turbine remain. There are early references to a watermill and a windmill belonging to Sullington Manor in 1297 and is probable that the present Chantry Mill stands on this site.

There are several other early references to Chantry Mill over the years, but care has to be taken as there were three watermills in Sullington parish. Edward Willmer was certainly occupying the mill in 1832 with William Hardwick in 1874, both also using the windmill on Sullington Heath (stopped 1907, burnt down 1911). It appears that over the ensuing years both the wind and watermill worked in conjunction until the windmill burnt down. In 1887, George Duke succeeded him and afterwards, Henry Crowhurst took over the ownership of the mill with his wife continuing until 1895 at least, employing John Quaite as the miller, but by 1905 Henry Barnes was in control until it closed in 1920. In 1921 the owner Mr Hicks removed all milling machinery and later installed a turbine, which generated Sullington's first electricity supply.

Chantry Mill is situated south-west of Storrington and sited on the River Stor which, even in the 1803 Defence Schedules, was barely sufficient to satisfy their regular customers. A millpond, with a high embankment was constructed to provide a sufficient head of water to power an overshot waterwheel, after which a turbine was installed to generate electricity, and although out of use in June 1947 it is still in-situ.

The mill has been converted into living accommodation and is built part brick, flint and timber to three floors under a half-hipped tiled roof.

GIBBON'S MILL *The Haven*
Tributary to River Arun – TQ 072 308 – At end of trackway west of Okehurst Road

Gibbon's Mill is not mentioned in the Domesday Survey as this part of West Sussex was once heavily wooded and mainly uninhabited. The first specific reference appears in the Nonae Roll in 1341, during the reign of Edward III. This document relates that amongst other things, the church had two watermills, Wanford Mill and Gibbon's Mill in the parish of Rudgwick.

Paul Adorian relates in his booklet *'The Story of Gibbon's Mill'* that in the Water Bailiff's book on the River Arun in 1636, no mention is made of any mill between Wanford Mill and Rowner Mill. The transcribed edition has written on the title page 'From E H Gibbon to his friend Chas Devon', which would appear to indicate that Mr Gibbon was in some way responsible for the transcribing, and perhaps his mill was omitted taking into account his responsibility for the repair of the bridge and highway - but this could be pure speculation!

In 1767, the mill was in the occupation of Harry Tullett and referred to as Gibbin's Mill and later as Gibbinges Mill. In 1812 Henry Botting was the miller with William taking over before 1845, and during his occupancy, the mill was almost doubled in size by a new brick and weatherboarded extension. This was probably built in anticipation of the Wey and Arun Canal being extended to Horsham, and although the existing canal was close by, the proposed branch would pass right by the mill. Botting, well aware of this and of the extra business it would generate, extended the mill, to provide more storage and associated ancillary machinery. The original waterwheel was retained and enclosed by a brick arch. The extension to the canal never materialised and Botting had left by 1869, possibly bankrupt. John Churchman, the neighbouring farmer, took over the operation of the mill and installed a steam engine in a shed adjoining the mill, to be operated when river levels dropped. During Churchman's ownership, Thomas Andrews and then James Harris were millers until commercial milling ceased at the turn of the century. A turbine that drove a dynamo and water pump in 1898 replaced the old undershot waterwheel. In 1930, a new 16hp Armfield turbine was fitted and in 1946, the dynamo and switchgear were replaced by a diesel engine. At this period of time the defunct mill provided a variety of uses as the turbine and the engine powered an 'Albion grinder, provided electricity, pumped water, and drove a grindstone, lathe and saw bench! The water pump ceased working when the main was laid to the farm and in 1960 electricity came to the farm and cottages.

Gibbon's Mill is a picturesque 4-storey brick and tarred weatherboarded building set in an idyllic location adjacent to a mill cottage of ancient construction. In 1993, the mill was being converted into a house and it was proposed that the turbine would once again generate electricity.

HEATH MILL *Pulborough*
Tributary to River Arun – TQ 073 176 – 1 mile southwest of West Chiltington

This was a rather small and insignificant watermill on the River Chilt, pulled down just after the end of WW1, of which little is known or recorded. Trade directories give William Heath, William Greenfield, William Botting and Reed & Stilwell as millers here during the 19th century, but nothing else is recorded. It was marked as 'Corn' on the 1896 Ordnance Survey 25" map, indicating that it was still in work, but only the small mill cottage, together with the remnants of the pond embankment, survives to indicate the site.

HORSHAM TOWN MILL *Horsham*
River Arun – TQ 163 303 – Set back from east side of Worhing Road

This was one of two large watermills situated close together on the south-west side of Horsham. The other, known as Stone Mill, operated by W. Prewett Ltd, was never water powered even though it was built in 1861. Prewett's took over Town Mill in about 1890 mainly for storage purposes, while Stone Mill continued produced flour. Although the Town Mill stands on an ancient site, the date of the first building is unknown. The Land Tax refers to a mill meadow in 1689, but it is first shown on a local 1734 map. In 1742 Resta and William Patching leased the mill and signed an indenture to serve the inhabitants of Horsham with pumped water from the mill, forcing it through wooden pipes to a reservoir in North Street. Resta Patching was declared bankrupt in 1745 and the business bought by James Andrews. He in turn sold it to John Smart, a millwright and John Mitchell a local plumber. In 1803, Mitchell sold his half of the business to Stringer Shepherd, and after his death in November 1824, it passed to his eldest son William. Later in 1844 William, the sole owner, sold the mill to Thomas Lee and then took out a lease on the mill at £144 and following his death in May 1852, his son Robert took over the running of the mill. In March 1860 the sale notice stated that 'it was in the occupation of Henry Allberry for a term of 14 years' and went on to describe the mill as 'having an overshot wheel and two flood wheels, a 8hp steam engine, 5-pairs of stones, flour dresser and bolter, millers cottage, 2 van sheds, piggeries, stabling for 3 horses and a cart house with a granary over it.'

Allberry continued as tenant, with his father operating Wanford Mill, near Rudgwick and according to the Land Tax assessments of 1866/7, he took over when his father died in 1861. It was at the end of Allberry's tenancy at the Town Mill that it was extensively rebuilt with a four storey building replacing the low two storey building. During the rebuilding, substantial iron columns were installed to support each floor all carrying the inscription 'W. Cooper - Millwright - Henfield – 1867'. Later it was owned and operated by George Sharp from 1871 after he left Warnham Mill in 1881. In August 1883, the mill, sold by auction, was purchased by William Prewett who was operating the adjacent roller mill and by the turn of the century it was primarily used for storage although a limited amount of grinding was carried out. Prewett died in 1913 with the business carried on by Fred and afterwards Charles. While stone ground flour milling carried on at Stone Mill until the 1970's, the Town Mill was falling into disrepair and was converted into a house in 1982 and then into offices in 1991.

All of the machinery has been removed but was of unusual operation as the 4-pairs of stones could be driven from a layshaft worked by an overshot waterwheel at one end, or by the breastshot flood wheel at the other with an extra pair of belt driven stones. The iron overshot waterwheel, located within the centre of the mill below ground level, was 12ft in diameter by 7ft wide with three sets of arms in a good condition. The iron breastshot flood wheel, with Poncelot buckets, was 12ft in diameter by 3ft 3in wide located at the northern end of the building adjacent to the overflow sluice. The flood wheel was fed from an open channel (now filled in) and the other wheel by an underground culvert. Although there is no defined millpond, the River Arun was widened upstream to allow a certain amount of storage. The mill is in a good condition now set amongst pleasant landscaped gardens.

HURSTON MILL *Storrington*
Tributary to River Arun – TQ 073 160 – North side of Hurston Lane

Very little is known about this mill demolished at the turn of the century. *Kelly's* 1866 directory records John Cootes as the occupier, succeeded by Richard & Edward Emery until 1899, after which the mill closed down although it was marked on the 1907 Ordnance Survey 6" map.

Of the mill site, situated by a sharp bend in the road near to Hurston Place, there are the remains of the wheelpit and the brick built pond wall, and according to Mr Hawksley it had two waterwheels.

KIRDFORD MILL *Kirdford*
Tributary to River Arun – TQ 001 262 – 2 miles southwest of the village

Contemporary photographs of Kirdford Mill, taken just before the turn of the century, show a well-established mill with a large breastshot waterwheel and mill house. However, a visit today reveals nothing apart from a mass of thick and tangled vegetation with the disappearance of Kirdford Mill complete.

It is not thought to be an ancient mill site, being first shown on Gream's 1795 map and was just south of 'Idehurst' a large house and estate.

The Sussex Marriage Licences of 1770 and 1773 refer to Edward Dale and Robert Osborne respectively, as millers at Kirdford. The mill is mentioned in the Defence Schedules of 1803 while later in 1827, according to the sale particulars in the Sussex Advertiser of 10 September, the watermill was let to William Court on a yearly basis. *Kelly's* trade directories record that John Baker was the occupier but, by 1890, he is just listed as a farmer and had connections with the watermills at Shillinglee and Ebernoe (there being no mention of the mill at Kirdford).

This was a small 2-storey mill with a stone base and haphazard weatherboarding, which had a waterwheel, whose diameter was out of all proportion to the mill. It is not known when the mill was demolished but in 1940, only odd pieces and brickwork remained. A track used to lead from the mill into the centre of the village, but even this has disappeared.

The site remains much as it did in 1940, with the leat to the mill still extant but that too almost engulfed in vegetation.

LINFOLD FARM MILL *Kirdford*
Tributary to River Arun – TQ 025 259 – East side of Kirdford to Strood Green Road

There is nothing to be seen and a modern house occupies the site, which lies one mile downstream of Kirdford Mill. There is a complete dearth of historical information about the mill apart from that it was built in 1818 and it could have been a farmer's own-use type. Outside the property is a millstone testifying to its former use.

NUTBOURNE UPPER MILL *Nutbourne*
Tributary to River Arun – TQ 077 187 – on side road east of The Street

This was a small watermill measuring only 15ft wide and 25ft long, which had a small overshot waterwheel driving 2-pairs of stones.

It is not clear whether this was an ancient site but it is more likely that this was the only mill to stand here dating from the late 18th century. It was one of two very small watermills that served the village of Nutbourne working in conjunction and according to a sale notice that appeared in the Sussex Advertiser in April 1835, both watermills were in the occupation of William Stovold, he being followed by Blaker Caffyn. After Caffyn, the Upper Mill was occupied by a succession of millers according to directories i.e. 1845-62 Henry Moase, 1860 Alfred Allen, 1870-74 Charles Clarke, 1878-82 Ezekiel Wicks, 1887-90, J. Goodacre (steam), 1890-99 Henry Stilwell (also at Lower Mill and Heath Mill). According to *Kelly's* 1938 directory 1938, Frank Terry, a local baker, owned the mill, but according to an inspection in January 1939, the waterwheel and machinery had been removed. Prior to its removal, the wheel had been briefly used for pumping water and for driving a dynamo.

It once had a large millpond, on the other side of the causeway, but this has dried up and the mill has been converted into a house.

NUTBOURNE LOWER MILL *Nutbourne*
Tributary to River Arun – TQ 077 185 – South of Upper Mill

The site is just below the Upper Mill, situated between two ponds, and had a succession of millers during the 19th century, until it stopped in about 1895, with Charlotte Reed the miller. The mill, built of dressed sandstone was finally pulled down in August 1937, having had a small overshot waterwheel 3ft wide.

In recent years, both ponds have been extensively regraded and landscaped but, at a point between the two ponds, a french burr stone, pieces of a peak stone, one half of a pit wheel and a cast iron stone nut can be seen. Access can be made from the Upper Mill or from a footpath west of the site.

OLD PLACE MILL *Pulborough*
Tributary to River Arun – TQ 045 191 – North of Coombelands Road

There is mention of a mill in 1500, which takes its name from the house almost opposite, built between 1404 and 1437. The mill was in the ownership of the Apsley family for many years and in 1796 a Mr Hammond had taken over. Thereafter, little is recorded, apart for directory entries i.e. 1838 - John Cootes, 1870 - Henry Pritchard. In about 1855, when John Stoveld rented the mill, the farmer objected to his plans to provided steam assistance, after which he left to take over Cripplegate windmill, near Southwater.

It appears that commercial milling ceased at the end of the 19th century, but no doubt the mill continued on a casual basis. In 1936 the mill was converted into a house, and the machinery discarded. The old clasp arm overshot waterwheel was retained on the north side, but all traces of this have disappeared.

This is an attractive looking building, separated from the millpond by an embankment that provided the necessary height to power the overshot waterwheel. The building is of dressed sandstone and in 1994 only some odd pieces of shafting remain to indicate the past use of the site.

RACKHAM MILL *Rackham*
Tributary to River Arun – TQ 046 142 – North of village

This small mill, picturesquely situated on the eastern edge of Amberley Wild Brooks, contains most of its machinery. It appears to be an ancient site with a reference to a watermill, with three acres, belonging to the Bishop of Chichester in 1369. *The Sussex Weekly Advertiser* of 12 July 1773 advertised the sale and reported that John Allen was leasing it for £13. Later in 1792, the same paper announces the sale of the mill and gives a Mr Burchell as the tenant. *Kelly's* 1874 directory lists Joseph Brown as occupier, while from 1882 until at least 1895, Thomas Ford was in control. He relinquished it to Samuel Smith in 1899, who stayed until 1918 after which Frederick Packham took over. A Mr Daughtrey was the last recorded occupier when the mill closed down in 1925. By 1946, it was reported that the building was disused and the new corrugated iron roof, erected in 1928, probably prevented the inside of the mill from deteriorating further. It was suggested that it continued using auxiliary power, but this was rejected. The disused engine shed dates back to when Joseph Brown was here; the engine being removed when Thomas Ford took over and converted it into a stable.

The external iron overshot waterwheel, 14ft 6in in diameter by 4ft wide, is reputed to have been installed in 1848 and cast at Hardham Foundry, but has few of its buckets left while the pentrough is supported on three brick piers. Inside, the 7ft 6in diameter pit wheel and similar size spur wheel, are cast iron. On the stone floor above, there are 2-pairs of 4ft stones, one peak one burr, set in octagonal wooden cases. The wooden crown wheel, 3ft 2in in diameter drove two ancillary pinions, the south of which drove a dresser, via a wooden pulley, while the north pinion powered the sack hoist which is set in a sliding frame.

The small millpond, fed via a leat from Parham Lake, has long gone and only a grassy embankment marks its position to the east of the mill. The mill is constructed in a mixture of brick, flint and dressed stone under a corrugated iron roof, with direct entry onto the stone floor, and there is evidence that the mill was extended some time in the 19th century.

ROWNER MILL *Billingshurst*
Tributary to River Arun – TQ 072 271 – 1 mile northwest of Billingshurst

This was an exceptional site, with both mills, built at different times, working directly off the River Arun. As the river and canal passed the site, the position of the two mills was critical as the area was prone to flooding and, to combat this, the mills were raised up on a small island using both low breastshot and undershot waterwheels.

The older of the two mills, according to an inspection of the mill in 1944, was built of timber and powered by a 12ft diameter low breastshot waterwheel that drove 3-pairs of stones, a wooden upright shaft, 7ft 6in diameter spur wheel and wooden stone nuts. Upstairs, there were three octagonal wooden tuns and the inscription 'Flood Nov 21,22,23 J. Sawyer 1815'. By contrast the newer mill, brick built with an iron undershot waterwheel, was about 18ft in diameter by 6ft wide but, by 1946, the machinery had been removed, with the doors and windows boarded up. There is a report that water pumping took place in this mill but little is known about this venture.

There is nothing to suggest that they were anything other than flour mills, as the 1896 Ordnance Survey 6" map refers to the mills as 'Corn', indicating that by then it was mainly just producing provender feed. During WW1 the mill apparently was restarted, for a short period, to produce cattle food.

The first reference is found on Budgen's 1724 map, while the first record appeared in February 1809 when William Carter insured his mill. According to *Pigot's* 1826 directory, James Carter was the occupier and it was still in the same family in 1842 according to the Tithe Apportionment, when reference is made to the second mill being built in the control of Joseph Carter. *Kelly's* 1855 directory names William Botting as the occupier, and he remained until 1870. It appears that Hammond and Son (later Hammond Bros) were the last occupiers until the mill closed down commercially towards the end of the 19th century.

An article in the *West Sussex Gazette South of England Advertiser* in October 1962 stated that the derelict mill was showing signs of water ingress. Both mills came into the possession of the West Sussex River Board in 1958 and despite last minute representations that they should be preserved, they were demolished in 1966. The area around the mill site floods on a regular basis and naturally the river board wanted to remove all possible obstructions to river flow. There is now nothing to be seen at the site, apart from the wheelpit of the older mill.

Rowner Farm is south of the mill site (on higher ground!) and in its garden is to be found an unusually constructed gazebo. A close inspection reveals a 7ft diameter spur wheel and a 5ft diameter crown wheel (from the older mill). A rather inglorious reminder to what was one of West Sussex's most attractive and picturesque watermills.

SHILLINGLEE PARK MILL *Northchapel*
Tributary to River Arun – SU 972 397 – 2 miles northeast of Northchapel

Beside the track on the approach to the mill, a large millpond (34 acres) indicates that the site was used in connection with the iron industry as, in the early 17th century, a furnace was operating here. This had shut down by 1620 and was replaced by a corn mill, shown on Budgen's 1724 map. As the mill formed part of a farm its principal use was as a corn mill (especially so in its later stages).

According to *Kelly's* 1866 directories, Matthew Taylor was in occupation and the mill stayed within the family until at least 1907. It worked in conjunction with Wassell Mill, at Ebernoe, between 1878 and 1907, but had shut in the early years of the 20th century. Afterwards a Gilkes turbine was fitted to power a pair of stones and later to generate electricity to Shillinglee House. The disused turbine still exists but the mill is devoid of any other machinery, and was converted into a house quite recently. The old mill is brick to 3-storeys, and at some time was later extended in size by a large brick built extension, which has a wooden lucomb.

SHUTT MILL *Bury*
Tributary to River Arun – TQ 007 150 – At Bury Farm

Little is known about this mill that was situated just to the west of the Petworth to Arundel road, but it appears to be an ancient site with a reference to it in 1199.

The mill worked in conjunction with a windmill at Cold Waltham according to a sale advertisement in the Sussex Weekly Advertiser of May 1809. Later the 1839 Tithe Apportionment refers to Charles Heather as the occupier and finally to James Hill according to *Kelly's* 1845 directory, which is the last reference to the mill although the site is marked as 'Corn Mill' on the 1876 Ordnance Survey 6" map.

The mill stood in the vicinity of Bury Mill Farm but is impossible to identify the site although an ornamental lake to the north of the farm buildings could once have been the millpond, but the disappearance of Shutt Mill is complete.

SLINFOLD MILL *Broadbridge Heath*
Tributary to River Arun – TQ 140 315 – End of track west of A264

A garage now occupies the site of Slinfold Mill but it appears that the lower brickwork was retained. This is an ancient site with the last mill constructed in the middle of the 18th century having two waterwheels located on opposite sides of the building.

Documents in the Horsham Museum referred to by George Coomber in his book *Bygone Corn Mills in the Horsham Area*, show that Henry Booker was the occupier of the newly constructed mill and continued until his death in 1787. Afterwards, the new tenant Edward Hersey was involved in an extraordinary catalogue of events with General Leland, the owner of the mill. After a series of claims and counterclaims about repayment of costs involved in repairing the mill, Hersey was declared bankrupt, but before proceedings could commence, he absconded from the mill and joined the army (he was later killed in Holland in January 1794).

It appears from the Land Tax Assessment, that Mr Acton was the occupier of the mill until 1797 after which Daniel Sharp took over and it remained under the control of the family until at least 1882. That it was a successful business was borne out as it was marked as a 'Flour Mill' on the 1896 Ordnance Survey 6" map, but it cannot have continued for many years after, as it was marked 'Disused' on the 1913 map, and partially demolished in July 1932.

The wheelpit for the undershot waterwheel, on the south west side of the mill, can be clearly seen along with the remains of a wooden sluice control, while the bricked up arch of the enclosed breastshot wheel is just visible. The mill house and cottage survive together, with the remains of a dried up millpond.

STORRINGTON MILL *Storrington*
Tributary to River Arun – TQ 087 145 - On northeast side of village

The first reference appears in a lease agreement between the owner, William Wheeler, and the tenant Thomas Coates in July 1731, and there is no suggestion that a previous mill stood on the site. Over the ensuing years it worked at times with the local post windmill and Chantry Mill. The mill was brick to three floors with a tiled half-hipped roof and had a large millpond.

According to *Pigot's* directories, Charles Wadlow was the occupier to 1843 at least, while Henry Joyes was here in 1870 and it stayed in the family until 1919 (with steam addition in 1911). The last occupants were T. Gatley & Sons, who were local farmers. In 1938, the external overshot iron waterwheel, 13ft in diameter by 5ft 6in wide, was reported in a good condition and the mill contained 3-pairs of stones (two in its latter working days) and iron machinery. Even as late as 1959, the waterwheel was driving, according to Denis Sanders, an oat crusher by a pinion off the crown wheel. During WW1 it was in full production and was known for the quality of its flour from 1-pair of stones and it worked occasionally until the early 1960's using waterpower.

The mill building lay disused for some time and there were proposals to preserve it, but ground conditions were bad and it was demolished. The penstock, manufactured by Blackaller of Steyning, along with other machinery, is now at Wateringbury Mill in Kent.

The pond has been restored and modern flats occupy the site and only part of an iron water feed pipe survives.

SWANBOURNE MILL *Arundel*
Tributary to River Arun – TQ 018 077 – South of Swanbourne Lake

This is a very old site, mentioned in the Domesday Survey and the mill, with the castle in the background, was immortalised by a John Constable painting completed on the 31 March 1837, the day before his death. The painting shows the mill as part of a large group of brick buildings standing on the edge of Swanbourne Lake.

On the 21 May 1774, John Tompkins the occupier took out nine oak trees to make a trough, penstock, waterwheel and axle, while later, according to a fire insurance policy in 1797, Robert Horne had taken over the mill. In January 1813, the *Sussex Weekly Advertiser* reported that he had been killed when his clothing was caught in machinery. *Pigot's* 1823 directory records Charles New as the occupier to at least 1839, with the mill closing down soon after, even though *Kelly's* directory later records George Bartlett as a corn merchant here, but he was only using the windmill. In 1844, a water pumping station was installed on the site of the mill, later powered by a water turbine (still in-situ).

Swanbourne Lake, is an artificially built millpond, now used for recreational purposes, but there is little to be seen at the mill site apart from an old brick arch and the faint indication of a tailrace.

WANFORD MILL *Bucks Green*
River Arun – TQ 085 327 – Southwest of Rudgwick

This is a medium size brick built watermill with timber infill, dating from the 17th century according to a recent structural inspection. At some time it was added to on its northern side, which included a new wheelpit.

This is an ancient site and could be one of the two mills accredited to Rudgwick in 1341, but the first definite reference appeared in a 1582 Deed for a new mill here. Apart from directories, little other documentary evidence survives, but in the 1803 Defence Schedules, the miller (unrecorded) could provide 23 sacks of flour as long as the wheat was supplied. According to the 1840 Tithe Apportionment, Henry Allberry was the miller and it remained in the same family until 1874, after which John Botting took over until WW1. Steam power had been added to the mill (from at least 1911) and after a short occupation by the Brewhurst Milling Co., Finskin Bros finally used the mill building as a store until just before WW2.

As with most mill sites, flooding was always a constant problem and when the derelict mill was being converted into a house in 1948, the following note was found. 'The highest flood ever known was in October 4/5th 1852 when water ran into the bottom floor of the mill, to a height of 14in, ruining about 200 sacks of wheat and other grain. Witnessed by Henry Allberry Jnr.'

It can seen from arches at the front of the mill that two waterwheels were used (each driving 2-pairs of stones), one internal wheel beneath the weatherboarded section and another within the brick built extension. The mill is of three storeys and devoid of machinery apart from some remains of the sack hoist.

WANTLEY MILL *Sullington*
Tributary to River Arun – TQ 089 160 – North of West Wantley Farm

Little is known about this site on the western outskirts of Storrington apart from the fact that it is marked on the Ordnance Survey first edition and in the 1083 Defence Schedules. Also, according to the 1842 Tithe Apportionment, William Heath was the occupier and Mrs J Standen the owner. Thereafter nothing more is recorded about the mill although it was thought to be of 18th century origin. There is nothing to be seen of the mill site.

WARNHAM MILL *Horsham*
Tributary to River Arun – TQ 168 323 – North side of Warnham Road

The watermill and large millpond are nearer to Horsham than the village from which its takes its name. This was the site of an iron furnace in 1609, ruined by 1664, with a corn mill afterwards to take advantage of the water supply. However, it appears that there was a previous corn mill here as according to the Warnham Church records in 1588, John Wellfare was the 'Myller of Warnham Myelle'.

In George Coomber's book, *Bygone* Corn Mills in the Horsham Area, he states that Henry Yates was the owner in 1700, with Thomas Dale as tenant (shortly to become owner) and he was still here in 1763 according to the Land Tax Assessment. Timothy Shelley of Field Place, father of the poet Percy Bysshe, had acquired the mill and enlarged his estate and the millpond at the turn of the 18th century. (Kenneth Reid, writing in the *Sussex County Magazine* of October 1948, is of the opinion that the mill was built in 1790, but this could be when the mill was modernised or rebuilt). By 1789 Thomas Ansell was the tenant until 1811, after which Benjamin Potter took over and was still here according to *Pigot's* 1839 directory, he being followed by his son Benjamin Jnr, after which George Sharp took over and stayed until 1881. He then left to take over the large Horsham Town Mill and William Prewett (later miller at The Town Mill and the steam mill in the Worthing Road), his assistant, took over the mill. In 1871 Charles Lucas of Warnham Court purchased the mill and in 1875 it was reported that Lucas wanted to pull the mill down. Fortunately this did not happen and instead it was refurbished and a new flour dresser installed. Kelly's directory shows Charles Crisford as the miller and then William Killick with Walter Cook the last miller, at which time, as a commercial concern, the mill closed. Over the ensuing years, the mill was used occasionally to grind corn for use on the home farm.

An inspection of the mill in May 1947 revealed that the mill was complete, although not in use, and was a small three storey building with brick upper storeys on a base of dressed stone, and although looking outwardly fairly modern, the roof timbers are extremely old. The report also stated that the overshot waterwheel 15ft by 4ft 3in wide, was cast iron with wooden buckets. The penstock is wooden with the water fed by a pipe and a small trade plate is inscribed 'Lintoft, Engineers, Horsham'. On the pit floor is a 10ft iron pit wheel, wooden upright shaft, 3ft wallower, and a spur wheel of most exceptional interest. At some time, the diameter of the pit wheel had been increased and the iron spur wheel had to be re-set in to the floorboard of the floor above with its rim 12in below the point of mounting. On the stone floor there were 3-pairs of stones in round wooden cases and a 4ft 6in diameter crown wheel which powered a flour dresser, other ancillary machinery and the sack hoist. The tail race is piped to a brick walled watercourse which passes under the Warnham Road before joining the River Arun between the Town Mill and Broadbridge Mill.

Simmons is of the opinion that the mill closed down completely as the varying water levels interfered with the skating of the pond, but there is evidence of a second waterwheel position on the east end of the mill that would have been overshot and about 4ft wide. In 1985 the mill was restored by E. Hole & Son of Burgess Hill and used as a antique shop and museum.

The building is used for office accommodation (1998) but most of the machinery survives and the waterwheel is turned twice a month. This is a very attractive red brick building with its distinctive Horsham slate roof tastefully restored, with the large millpond (now part of a nature reserve) still in existence.

WASSELL MILL *Ebernoe*
Tributary to River Arun – SU 901 281 – South of Steel's Lane

It seems that Ebernoe and its mill have a reputation for remoteness. This is another corn mill on the site of a forge, in existence in 1574, with sows of iron brought over from Shillinglee Mill until closure in 1641. According to the 1847 Tithe Apportionment, Clara Baker was the occupier and *Kelly's* 1878 directory lists Matthew Taylor with the mill working in conjunction with Shillinglee Mill until 1907. James Lusted took over until 1922, after which James Gwillim took control, running Coultershaw Mill, Midhurst North Mill and Fittleworth Mill at the same time. However, Gwillim's stay was short lived and by 1927, Holden Bros were in control. In 1948 the mill was converted into a house with the machinery removed, but the waterwheel retained. This wheel, cut in half and used for generating electricity is now dilapidated, but when complete, it measured 10ft 6in in diameter by 8ft wide set on a square iron axle shaft.

This is a picturesque three-storey mill of red brick and weatherboarding, set into the embankment that carries Streel's Lane on the opposite side of which, is the dried up millpond.

Birchenbridge Mill prior to demolition in 1958

Bignor Mill as a house in 1957

The rebuilt Brewhurst Mill at the turn of the century (RP)

A close up of the floodwheel at Brewhurst Mill in 1957

Mr Kitchener, the miller at Brewhurst Mill in 1956

Chantry Mill and pond in 1907 (VU)

The attractive Gibbons Mill in the early 1900's (RP)

The low two storey Horsham Town Mill in 1849 (AS)

Horsham Town Mill in 1905 (VU)

Conversion of Horsham Town Mill to a house in 1982 (NC)

Old Place Mill, as a house conversion in 1955

The miller John Baker at Kirdford Mill in 1895

The unusually large waterwheel at Kirdford Mill in 1895

Upper Nutbourne Mill and pond in 1910 (AB)

Upper Nutbourne Mill Lying disused in 1955

Remains of the waterwheel at Wassell Mill in 1994

Wanford Mill in 1957 showing two waterwheel inlets

Stone floor entrance to Rackham Mill in 1995

The remains of the waterwheel at Rackham Mill in 1995

Wassell Mill in the 1930s lying disused

Rowner Mill lying disused in 1956

The cast iron waterwheel at Rowner Mill in 1964

Wooden waterwheel at Rowner Mill in 1956

Shillinglee Mill as a house conversion in 1994

Storrington Mill working in 1946

Storrington Mill lying partly disused in 1955

A painting of Swanbourne Mill at Arundel (VU)

Warnham Mill in 1998 showing its earlier conversion

Warnham Mill in a ruinous state in 1956

RIVER ADUR

RIVER ADUR

ASHINGTON MILL *Ashington*
Tributary to River Adur – TQ 130 157 – At the end of Mill Lane

According to the *Victoria County History*, this site was established in 1538 and it later became a large commercial mill with two waterwheels working in tandem. In a sale advertisement in the *Sussex Advertiser* of 12 June 1837, it was listed as having two waterwheels, 4-pairs of stones and also included the Rock and High Salvington windmills. *Kelly's* 1851 directory records Henry Hampton here and in *Melville's* 1858 directory, his widow only, she being followed by Henry Humphrey, who continued until at least 1887. After short tenancies by Alfred Coote (1890) and Joseph Hammond (1895), John and Charles Muddle took over and it remains in the family to the present day.
Ashington Mill was an attractive building of 3 floors but a modern warehouse has replaced the picturesque wooden outbuildings. The two overshot waterwheels were removed in 1915, and replaced by a turbine. An inspection in May 1939, revealed that a portable 8hp Clayton & Shuttleworth engine was lying disused and the mill was worked by water. The mill was just producing animal feed but this discontinued soon after. The mill burnt down in 1974 and the remainder lies hidden within the large warehouse that continues to supply animal feed and other agricultural products. The once expansive millpond has disappeared without trace.

BOLNEY MILL *Bolney*
River Adur – TQ 263 218 – East of Garston's Farm

The mill was situated on the south side of a path that runs between Garston's Farm and the A23, but it was only erected at the beginning of the 19th century, according to a sale advertisement that appeared in the *Sussex Weekly Advertiser* of 15 February 1813:
"To be sold by auction on March 2, 1813 at the Castle Inn, Twineham. All that valuable freehold and leasehold premises, situate at Bolney, comprising a freehold watermill, newly erected on an improved plan, and fed by a very large piece of water, with a continual stream, and a leasehold windmill, situate on Bolney Common. The mills are nearly close to the new turnpike road from Brighton to London. Immediate possession may be had, and particulars may be known by applying to Mr P Barber, the proprietor on the premises".
There was a succession of different miller's here in the 19th century according to directories ie. 1842-58 Henry Leppard *(Melville's)*, 1866 Henry Payne, 1870-74, Thomas Ashby, 1878-82-Packham and Coomber *(all Kelly's)* after which all entries for the mill cease.
Bolney Mill stood in an isolated position and two detailed inspections of the mill in 1939 and 1946, gave an insight into its contents and construction. It was a typical 3-storey brick and tarred weatherboarded building measuring only 17ft 6in long by 16ft wide, with a prominent mansard roof. There was a small wooden addition on the stone floor with an open shed below. The mill was set below the pond embankment with entry directly onto the stone floor. It had a large external iron overshot waterwheel, 17ft in diameter by 4ft wide, bearing the inscription 'S. Medhurst & Son Sept 1861', while the penstock bears the inscription 'J. W. Holloway, Engineer, Shoreham Sussex'

Inside, the machinery was cast iron and of an age commensurate with that of the waterwheel. The pit wheel and spur wheels were unusually large, being 12ft and 10ft diameter respectively, but there was no crown wheel as the upright shaft terminated at the pit room ceiling.

Just after the last war, the cogs were stripped from the spur wheel and only an oat crusher was being worked off bevel gears, but soon after, the mill was out of use and it never worked again. It was demolished in April 1964, according to Frank Gregory, and was bulldozed to the ground and then buried under an earth mound. Bolney Mill was a small country watermill and apart from its own millpond, a large pond upstream acted as a reservoir.

BOLNEY WEST MILL *Bolney*
River Adur – TQ 250 233 – Near Colwood Court

The mill is marked on Budgen's 1724 map while the 1879 Ordnance Survey 6" map refers to the site as 'Old Mill Farm', (which has since become Colwood Farm). It was presumed to be an iron working site but no documentary evidence supports this, but the *Victoria County History* states that gunpowder was made here, but again no other sources corroborate this, and there is nothing to be seen today.

COBB'S MILL *Hurstpierpoint*
River Adur – TQ 273 189 – North end of Langton Lane

Cobb's Mill is a brick and timber building with a large mill house on its northern side, situated on the upper reaches of the River Adur. It is one of the few Sussex mills that could be restarted, as it only stopped working in 1966. This does not appear to be an ancient site as it is first referred to on Budgen's 1724 map and after a partial rebuild in 1869, most of the machinery was upgraded.

The mill was occupied of Anthony Ede according to the 1803 Defence Schedules but in *Pigot's* 1839 directory, Charles Packham was the tenant and the mill remained with the same family until 1938. Between 1919 until 1938 the Packham's were also using the large Ruckford Mill, a mile upstream, both run under the name of Charles Packham Ltd.

The iron overshot waterwheel, on the south side of the mill, measures 11ft in diameter by 6ft 10in wide on a square iron axle, although its 48 buckets were removed prior to restoration, which never took place. A cast iron pentrough, fed by twin iron pipes above the wheel, is inscribed 'H. Cooper, 1868'. The machinery is comparatively modern with the 11ft diameter pit wheel only dating from 1962, while the cast iron layshaft latterly drove 4-pairs of millstones, 3 composition and one french burr. Pride of place in the mill is the 50hp Tangye single cylinder suction gas engine, installed in 1906, which ran in association with the waterwheel. The mill had its own gas producing plant situated in a brick building behind and above the mill, beside the attractive millpond.

Following the closure of the mill, the buildings were kept in a good condition, but have rather deteriorated in recent years. The owner is however, anxious to preserve this interesting and important West Sussex watermill.

COPSALE MILL *Copsale*
Tributary to River Adur – TQ 169 249 – Adjacent to Bar Lane

This is probably a fairly ancient site as there are references to a mill here in 1502, but over the ensuing centuries, little is recorded. It is marked 'Flour Mill' on the 1879 Ordnance Survey 25" map and 'Disused' on a later map of 1912. *Kelly's* 1887 and 1890 directories refer to John Hide as the miller but nothing else is known about the mill itself. According to the *Victoria County History*, the mill was disused by 1896 and demolished by 1939.

The site is situated to the west of the old railway bridge at Copsale and is just discernible by a dried up tumbling bay, while on the other side of the road is the vague outline of the former millpond.

COURT MILL *Steyning*
Tributary to River Adur – TQ 171 115 – Off Mill Lane in Steyning

Court Mill is situated on the western outskirts of Steyning, which, during the Domesday Survey, was an important town with its own mint and market. A watermill was in existence here before the conquest, with the last mill finally stopping in 1927, having previously worked in conjunction with Gatewick Mill, the other Steyning mill, and the windmill at Round Hill.

The 1803 Defence Schedules mentions the mill but not the miller and it is not until an account of a robbery here, in the *Sussex Weekly Advertiser* of November 1816, that the miller Peter Lashmar is named, with his son Thomas carrying on the business until at least 1845. The next occupier was William Penfold, who was a well known local character who was never seen anywhere outside without a hoe or pitchfork, or some long handled implement, which he used to swing about whilst walking, or lean upon when talking. Apparently, his method of bringing a conversation to a close was to pull out his watch, glance at it and remark " - o'clock and nothing done and we shall be in the workhouse".

By 1878, Penfold was in partnership with Thomas Brown, the latter running the mill on his own with his wife the occupier from 1890 until 1899. It is certain that commercial flour milling had ceased towards the end of the century even though a steam engine, with an enormous boiler chimney, had been added. Thereafter, the mill carried on grinding animal feed and making dog biscuits until its final closure in 1927, when in the ownership of the Steyne Food Co.

Within three years of closing, the machinery had been dismantled and removed, with the inevitable conversion to a house. The old waterwheel and some of the pit machinery was retained but later on a new waterwheel was fitted for aesthetic reasons. The overshot waterwheel, 17ft 6in in diameter by 4ft 6in wide, is constructed in iron with wooden buckets on a now rotten wooden axle shaft. Tradition has it that when this wheel was being fitted the date '1650' was found scratched on the axle shaft. C.A. Woolgar made the waterwheel at his wheelwright's shop in School Lane, and to make certain that everything was in order, he assembled the waterwheel on the upper floor of his workshop before erecting it. The other feature of interest is the cast iron metal pentrough with the inscription 'W. Cooper. Millwright and Engineer Henfield 1872'.

The mill, and the attached mill house, forms a most attractive group of buildings with the brick and timbered mill of greater age than the house. Inside, some of the pit machinery survives, including the 10ft diameter spur wheel, but the upright shaft has been removed to another part of the mill house. Outside, the mill is set partly below the millpond embankment.

CUCKFIELD UPPER MILL *Cuckfield*
Tributary to River Adur – TQ 292 241 – West of A272 southwest of Cuckfield

This is certainly an old site that stopped working during the middle of the 19th century. In the

Sussex Weekly Advertiser of 12 September 1803, it is reported that Henry Jeffery, a miller at Cuckfield (working for Caffyn) handed over his estate and effects for the benefit of his creditors. This is the last reference to the mill and Highbridge Mill, on the opposite side of the A272, took over.

The mill was leased to Thomas Caffyn in 1799 and the same family were working Highbridge Mill during most of the 19th century.

An inspection in June 1947, indicated that the site was marked by a waterfall, comprising of stone blocks some 100 yards from the mill house, while a visit in 1994 revealed just a small grass mound which could have been part of the millpond embankment. The mill house is of considerable age and is reached by a long and private winding track from the A272.

GATEWICK MILL *Steyning*
Tributary to River Adur – TQ 178 113 - Between Church and Gatewick House

Gatewick Mill was demolished in 1878, and a contemporary photograph shows it to be a small wooden building of great age. The mill worked in conjunction with Court Mill and the windmill at Round Hill, certainly after 1839 when in the control of Thomas Lashmar. It was marked 'Flour' on the 1879 Ordnance Survey 6" map (surveyed before the mill was pulled down) and was fed by a small stream that ran from the millpond at Court Mill. The mill was situated adjacent to Church Lane near St.Andrew's Church and Gatewick House, and only a small grass mound marks the site.

GOSDEN LOWER MILL *Lower Beeding*
Tributary to River Adur – TQ 228 248 – Perryfield Lane

The *Victoria County History* records that the mill was either being built or rebuilt between 1597 and 1605, but George Coomber writing in *Bygone Corn Mills in the Horsham Area* states that a mill was first mentioned in the Beeding Manor Accounts of 1400 & 1439.

This was one of three mills in close proximity to each other (one being a windmill) that often worked in conjunction. A sale notice in the *Sussex Weekly Advertiser* of 11 July 1774 stated that Joseph Terrell was the occupier, but in 1779, he had left and was replaced by James Souch (the new owner) who stayed until 1791. Souch let the mill to Richard Terrell for 20 years and then William Rayward, who left in 1850. The mill is marked as 'Flour' on the 1874 Ordnance Survey 6" map and it continued as such until the 1890's.

The site lies on the west side of Mill Lane by a sharp bend in the road, and a report in November 1938 points out that the remains of a black square feed pipe (which used to convey water under the road) lay abandoned on site. By 1992, little has changed apart from the site becoming overgrown, but a close inspection reveals the remains of a wooden sluice gate on the opposite side of the road, but the small millpond has all but disappeared.

GOESDEN UPPER MILL *Lower Beeding*
Tributary to River Adur – TQ 229 249 – East of Mill Lane

This is the second of the two watermills near Mill Lane that worked in conjunction with the Lower Mill. It was situated at the south end of a pond which forms part of Leonardslee Gardens, while the pond further upstream worked an iron furnace.

According to *Kelly's* 1878 directory, Amos Killick was the occupier, with Mary Killick the last recorded occupier in 1898, but the Ordnance Survey map reveals that it was disused at the time. The Killick family moved to new premises in Mill Lane which were driven by an oil engine, but on the 30 October 1933 is was destroyed by fire.

There are remains of stone walling, set below the pond, at the watermill site, reached by a footpath leading off Mill Lane.

HAMMOND'S MILL *Clayton*
Tributary to River Adur – TQ 300 176 – West of Hassocks to Burgess Hill Road

Hammond's Mill has gone and there are few remains to indicate its exact postion. This is an ancient site as the first reference appears in 1482. According to the Land Tax Returns, Mrs Owen was the owner in 1780; her tenant being John Godley, who remained under different owners until 1810. John Hamlin took over the tenancy, and was followed by William Osborne in 1813, while in 1817, John Gainsford was the occupier and a Richard Gainsford appears as the owner in the following year, and continued until 1832. The 1844 Tithe Apportionment records John Mercer who was here to at least 1862, according to *Kelly's* directory. William Wood who, in 1882, was working the nearby Ruckford Mill in conjunction, followed him. In 1895 the mill was operating under the name 'New Close Flour Mills' (New Close Farm being by the main road), and run by mill managers until 1903 and in a sale notice, the mill was said to contain 3-pairs of wheat stones and to be in excellent running order with a turnover of 100 sacks of flour per week. After flour milling ceased in 1903, Sydney Simmons is of the opinion that the building was used for the production of copper wire, but nothing is known about this venture.

Hammond's Mill ended its working days occasionally grinding animal feed and was demolished in 1975. It was a square building with three floors of brick and one of timber, with the pit floor below ground floor level. It was standing disused in June 1946 when an inspection was carried out, which stated that there was plenty of storage space in the mill with the upper floors described as 'very roomy'. Further storage space was available in the extension on its southern side, and circular steel columns supported all floors, with a brick built engine house on its western side. The iron overshot waterwheel, 11ft in diameter by 6ft wide, was enclosed at its eastern end with the inscription 'W. Cooper, Millwright, 1870' on the rim of the wheel, with the drive machinery of a date commensurate with that on the waterwheel. The mill had 3-pairs of stones, although one pair had been removed, while on the stone floor there was an ancillary drive to machinery in an adjacent barn. The pentrough was iron and fed by a pipe that was also enclosed, with its upper part protruding through the stone floor. The pentrough was inscribed 'W. Cooper, Millwright and Engineer, Henfield 1871'. Water was received from a small millpond on the opposite side of the track that passes the mill site.

The last mill was built in 1821, according to a stone tablet, while the wooden top floor was added in 1880 and the mill cottage has, on its chimney stack, a tablet bearing the date 1743. This is now a depressing site for, before its demolition, it was in a fairly good condition and efforts to preserve the waterwheel at least, were prevented when it was literally smashed to pieces, but the iron pentrough was rescued intact and has since been installed at Ifield Mill.

HIGHBRIDGE MILL *Cuckfield*
Tributary to River Adur – TQ 297 237 – East of Ansty to Cuckfield road

During the 19th century at least, Highbridge Mill worked in conjunction with Cuckfield Upper

Mill, on the opposite side of the A272, at a time when the Caffyn family were operating both mills. The family connection started when Ann Sergison, of Cuckfield Place, leased two watermills to Jacob Caffyn in 1799 and, according to directories, the Caffyn's worked the mill here to at least 1927. F.Dann & Son took over and continued until at least 1938 after which, all directory entries cease. According to an inspection of the mill in 1946, it was only in very occasional use and a year later, was bought by Jenner & Higgs of Bridger's Mill and closed down.

Highbridge Mill is a small brick built building set into the side of a large pond embankment that almost dwarfs the height of the mill. Before additional residential accommodation was built here, there was a small wooden storage building in front of the mill. The mill is built to three floors with red brick with a weatherboarded gable under a tiled mansard roof, with the extant cast iron overshot waterwheel positioned on the north side, although it is now without any buckets. It is 13ft in diameter by 4ft 6in wide and in reasonably good structural condition. Although there is no pentrough, a gate box fed water onto the wheel through a 12in diameter steel pipe. The pipe that took water from the millpond now lies abandoned and acts as a barrier to stop cars reversing from the car park into the millpond! The pit machinery remains virtually intact and is of a conventional layout, with the 10ft diameter pit wheel connected to the wooden upright shaft by a very rare wooden wallower. The 9ft diameter wooden spur wheel once had a ring drive for engine assistance. This was a 2-pair mill, both of which remain on the floor above, while the 5ft diameter cast iron crown wheel is complete with new wooden teeth, that were put in when work on the repair of the mill was briefly carried out in 1960. There are ancillary attachments from the crown wheel that was once were connected to the sack hoist and other equipment. The bin floor has been converted to office accommodation.

One of the outstanding features of this site is the millpond excavated in 1810. This was drained when a previous owner demolished the sluice gates, but the depth and extent of the millpond would have ensured a sufficient capacity to support a days milling. Recent building work has deprived the mill of any apparent associations with its past when viewed from the access track, but at the back, the mill still stands and with the nearby mill cottages, forms a most attractive location.

HOOKER'S MILL *Twineham*
Tributary to River Adur – TQ 256 204 – East side of road at Twineham Green

This was certainly a most unusual mill that, over a short period of time, was powered by water, wind and finally steam!

The mill was erected in 1851 and originally water powered, but this did not prove a success due to problems with the water supply filling a small millpond. A windmill was then constructed on a flat roof extension to the watermill, and used to supplement the water power. Finally, both methods were abandoned and the mill was driven by steam, but even this could not provide a viable business venture and it closed down in 1890.

Soon after it was built, the mill was owned by John Wood in 1859 and he leased the mill initially to Arthur Thompson and then to Gorringe & Son until 1875. Afterwards, it came under the control on the Packham family, who were also operating Bolney Mill, Ruckford Mill, Cobb's Mill and Hurst mill near Petersfield. It was George Packham who erected the windmill in 1876, with Charles Packham the last tenant. The windmill was damaged in 1887 and pulled down soon after, the watermill machinery having been removed previously. The windmill was a four posted open trestle mill, similar in design to that which stood at West Ashling Mill, near Chichester.

A lease, advertised in *The Miller* in February 1876, indicated that the mill had 5-pairs of stones and apart from the mill house, there was a blacksmith's and a bakery on the site. Ernest Hole, the well known Sussex millwright, was employed to dismantle the remaining machinery and demolished the buildings in 1900.

The site of the mill can still be identified at Twineham Green, on the opposite side of the road from Hooker's Farm, from which it took its name. There are pieces of odd brickwork and a dried up mill race, but nothing else to indicate the size and magnitude of this long forgotten watermill.

KEYMER MILL *Keymer*
Tributary to River Adur – TQ 314 140 – South of village west of Lodge Lane

This site is shrouded in mystery and is only referred to on Budgen's 1724 map, but comprehensive landscaping has taken place, making it difficult to identify the site. There are two ponds, the lower of which could relate to the mill site, as there are traces of very old brickwork and the faint outline of a bypass channel on the edge of woodland.

KNEPP MILL *Shipley*
Tributary to River Adur – TQ 162 212 – Beside private access to Knepp Castle

Knepp Mill has disappeared and become one of the many lost mills of Sussex, the large pond first served an iron furnace, belonging to the Duke of Norfolk.

The watermill was indicated on a 1754 estate map, and later shown on Gream's 1795 map. Knepp Castle was built in 1806 and partially burnt down in 1904, and subsequently restored, while the remains of the mound of the old Knepp Castle, one of a line of Norman garrison posts, can still be seen just to the south. A disused pumping house, half buried, is to the north-west, but little evidence of the mill survives, apart from the slight indication of a leat leading from the pond in the direction of the half timbered mill house, which could have once been the mill.

The large pond was constructed for the iron furnace situated almost adjacent to the A24, where a group of very old buildings could date from this time.

LEIGH MILL *Hurstpierpoint*
Tributary to River Adur – TQ 287 212 – South of Ansty

This does not appear to be an ancient site and a date stone on the original mill 'CP 1808' suggests when the mill was established. The mill was eventually made up of two buildings, the older being brick, timber under a mansard roof adjacent to the millpond, whilst the extension of 4 floors with a flat roof, was a most unattractive building used for storage.

From 1878, and probably earlier, it was run by the Packham family who were, at the same time running other watermills in the locality, notably Cobb's Mill. Benjamin Packham was the first miller, he being followed by James, who ran the mill until flour milling stopped in 1913-14. The mill was then purchased by Major Saunders, who continued running the mill until about 1933 for chaff cutting and oat crushing, but afterwards it never worked again.

An inspection of the mill in 1946 provides an insight into its method of operation. At the time the mill was complete, but dilapidated, with the 12ft diameter by 5ft wide overshot waterwheel in a very poor condition. The machinery was predominately wood although the 8ft diameter pit wheel, axle and bridge trees were of iron. The upright shaft, the 8ft diameter spur wheel, and the 4ft diameter crown wheel (all wooden) powered 3-pairs of stones, 2 peak, 1 burr in octagonal wooden cases. The waterwheel and pentrough were manufactured by Coopers of Henfield.

After lying disused for some time, the mill was finally demolished in the 1950's. A visit in 1991 revealed the remains of the brick base of the original mill, along with the remains of the pit wheel, while the millstones were removed to the Jill windmill at Clayton.

MANOR MILL *Poynings*
Tributary to River Adur – TQ 262 123 – West side of road leading north from village

The small village of Poynings, tucked away behind the South Downs, had two watermills, with Manor Mill the better known. The village dates from early times and it is assumed that a watermill had been in existence, certainly since 1339, when the manor possessed a mill.

The existing mill house has a date stone inscribed '1625' and Simmons concluded in 1939, that it was by then one of the oldest in Sussex. In the 1803 Defence Schedules, the miller could provide 30 sacks of flour every 24 hours (with just 2-pairs of stones and an irregular water supply this must have been some feat, and is probably wrong.) According to the 1843 Tithe Apportionment, James Grammes was the occupier, he then passed control over to Timothy Grammes, who ran the mill until at least 1870. After a short occupancy by James Longley, Charles Tulley took over and ran it in conjunction with Mill Farm. It continued working commercially until 1895, after which all references cease, but no doubt it continued as a farm mill for some time after, but in 1939 it was in a very poor condition. The mill was a comparatively small building comprising two floors of tarred weatherboarding over a ground floor of brick, then in use as a household store, its timbers for the most part rotten. The mill could be entered directly into the stone floor that contained 2-pairs of stones and a solid wooden clasp arm crown wheel set on a modern iron upright shaft. The line shaft, from the crown wheel, engaged a pulley that protruded from the mill, and could be belt driven from a small steam engine housed in an outbuilding (now demolished).

In 1946, the mill was demolished with the mill house repaired on its southern end. A contemporary photograph of the mill, taken at the turn of the century, shows it to be in reasonably good condition, apart from the waterwheel, which had been rather haphazardly strengthened by the use of wooden timbers attached between each arm.

Although the mill was demolished over fifty years ago, the remains of the waterwheel are to be seen together with a half buried iron pit wheel (approx 7ft in diameter) and the wallower. The millpond has dried up and now just the attractive mill house remains to perpetuate the name of Manor Mill.

OLD SHOREHAM MILL *Shoreham*
Tributary to River Adur – TQ 215 057 – West side of Mill Lane

There is a dearth of information about this mill that, according to an old 1850 map of Shoreham, was on a small tributary that issued into the old Shoreham Harbour. The map shows a millpond behind the mill, situated beside Mill Lane. According to Frank Gregory, it was a small corn mill converted into a house before 1937 (it has since been demolished).

It is impossible to identify the site today, although a small embankment south of the cemetery lodge could have formed part of the millpond.

RUCKFORD MILL *Hurstpierpoint*
River Adur – TQ 293 180 – Northeast of Hurstpierpoint College

This brick built mill is situated on the upper reaches of the River Adur between Cobb's Mill and the site of Hammonds Mill. It is a large building containing most of its machinery and two waterwheels. It is not clear how old the site is; it being first shown on Budgen's 1724 map and it would appear that the existing building would date from about then.

After being named as 'Avery's Mill', it is shown on the 1843 Tithe Map as Rickford Mill, but in the latter half of the 19th century it became Ruckford Mill. The 1803 Defence Schedules record George Bishop here and he could supply five sacks of flour every 24 hours. From 1855, William Wood was the occupier and stayed until 1909 (he was also using Clayton windmill in 1905), while *Kelly's* directories record that from 1919 until 1938, Charles Packham Ltd were owners and ran the mill in conjunction with Cobb's Mill.

There are two waterwheels enclosed at the northern end of the building, side by side with one half in front of the other and both overshot waterwheels are 12ft in diameter. The outer wheel, 4ft 3in wide, has an iron rim with a wooden sole and wooden arms (renewed in 1937 by West of Burgess Hill) with the 16in diameter octagonal wooden axle shaft passing in front and close to the inner wheel, which is 4ft 9in wide. The top of both wheels almost touch the ceiling of the wheelhouse, with the inner wheel fed by two pipes and the outer by a single pipe. In 1940, the iron wheel was reported to be in an excellent condition but apparently wasted a lot of water. The rack and pinion gate controls were operated from inside the mill, with the outer wheel control inscribed 'W. Cooper, Millwright and Engineer, Henfield 1870', and the inner wheel control inscribed 'W. Cooper, Millwright 1861'. Most of the machinery is iron and is as follows:
Outer Wheel: 7ft 6in pit wheel, 3ft Wallower, 7ft 6in wooden clasp arm spur wheel, drive for one pair of stones, iron upright shaft, 4ft 6in crown wheel with pinions to oat crusher and sack hoist. Inner Wheel: 8ft pit wheel, iron upright shaft, 8ft 6in spur wheel, (originally wooden but replaced in the 1930's), drive to 2-pairs of stones, 6ft crown wheel with underside teeth. (The drive from the oil engine connected with this crown wheel).The engine house still survives and has been converted into a kitchen.

According to Denis Sanders, who inspected the mill in July 1959, it was producing animal food, powered by the oil engine. It is also recorded that the mill closed down in January 1966, and its solid construction lent its conversion to a house soon after. The mill is built to four floors, under a tiled roof, with a large 2-storey lucomb projecting over the front of the mill.

This is a typically attractive site with the old mill kept in a good condition along with its machinery, although both wooden axle shafts are in a poor condition.

SHERMANBURY MILL *Shermanbury*
River Adur – TQ 213 187- Southwest corner of Shermanbury Park

There is now little to be seen, apart for some brick footings and traces of the dried up water channels. The mill was of three floors, two of brick and one tile hung, and although not a large building, it had two waterwheels at opposite ends.

This is an ancient site dating back to the Domesday Survey, while in 1611, John Gratwicke, lord of the manor, was apparently running the mill himself. By 1745, Henry Francombe was owner of the estate and carried out extensive improvements to the floodgates and to the main river itself. The mill was occupied by the Ede family from at least 1777, when William Ede advertised for a journeyman miller according to an advertisement in the *Sussex Weekly Advertiser* in March of that year. He was still here when the last mill was erected here in 1816.

Directories record that a succession of millers occupied the mill after William Ede during the 19th century, i.e. 1826 - Henry Stabler, 1832-39 - John Burtenshaw, 1862 J. Fuller - 1878, Albert Morgan and finally 1882 - S. Copestake. It was marked 'Flour' on the 1879 Ordnance Survey 6in map, but did not continue as a corn mill as it became a saw mill thereafter.

This was not a particularly large mill and its two waterwheels only powered 4-pairs of stones between them. The breastshot waterwheel on the east side of the mill was 13ft in diameter by 10ft 8in wide and was a fine example of its type. The waterwheel on the west side was smaller and the last to be used, it being 11ft 6in in diameter by 3ft wide. Both wheels powered wooden upright shafts and drove the following machinery, East Side: 11ft 6in pit wheel, 3ft wallower, 7ft clasp-arm wooden spur wheel, two millstones, one 4ft peak and 3ft 8in burr, 4ft crown wheel. West Side: 8ft pit wheel, 3ft 6in wallower, 6ft 6in clasp-arm wooden spur wheel, 4ft crown wheel.

According to an inspection of the mill by Sydney Simmons in June 1946, the layout of the shafting, adapted to drive the saw bench, was very unusual. A 12in and 14in bevel from the south east side of the west crown wheel and the south west side of the east crown wheel respectively, drove a rather complicated system of shafting to the saw bench outside the mill. A Victorian two-storey extension was built on the west side for living accommodation and the tail water from this waterwheel was conveyed in a culvert under the building.

The mill remained disused for many years and suffered badly when troops were camped in the park during WW2, and was later demolished. It stood in the south west corner of Shermanbury Park, near to the church, with the water supply provided by a large ornamental lake which was fed by a small stream from the River Adur.

SLAUGHTER BRIDGE MILL *Coolham*
Tributary to River Adur – TQ 122 234 – West of Coolham Road

The mill takes it name from the adjacent Slaughter Bridge, north of the village of Coolham. It was built in about 1780 on behalf of the Killick family and was a small brick and stone built mill of two floors with a small millpond, with an undershot waterwheel and a lucomb later covered in sheet metal.

Henry Killick was the recorded occupier until 1859, after which E.Joyes was the tenant, he being followed by James Thorpe until about 1888, when the 1896 Ordnance Survey map describes it as 'Corn'. Little else is known about the mill except that it was occasionally tractor driven after commercial milling ceased. According to a report, the mill was still standing in 1957, but was demolished a few years later. A petroleum depot occupies the site and all traces of the mill have disappeared.

SPRING MILL *Poynings*
Tributary to River Adur – TQ 264 118 – South side of village

The site of Spring Mill is situated on the slopes of Devil's Dyke, one of the best known landmarks in the South of England, but an article in the *Sussex Archaeological Collections*, written in 1862, stated that Spring Mill was newly built.

This was a small farmers own use mill, housed in a barn powered by a spring that formed two ponds, the lower of which being used directly for the mill. It seems that the mill was destroyed by fire in the 1870's when John Williams was the occupier, he having taken over from William Botting. It was reported that after the fire, the stones were taken out and the machinery sold for scrap.

Nothing remains apart from the retaining walls of a wheelpit, indicating a narrow waterwheel, and a much reduced millpond, reached by a footpath from the village.

TRUSLER'S MILL *Albourne*
Tributary to River Adur – TQ 249 165 – East side of Trusler's Hill

A mill is mentioned in 1615, according to the *Victoria County History*, belonging to the Albourne Manor estate, and although marked 'Albourne Mill' on the 1879 Ordnance Survey 25" map, it was thereafter known as 'Trusler's Mill'.

According to the *London Gazette* of April 8 1815, bankruptcy proceedings were taken out against James Burtenshaw, and Thomas Wickham became the new tenant and it remained in his family until at least 1928. This was a small brick and timber 2-pair mill, operated by a small head of water to the east of Albourne opposite Trusler's Mill Farm, that was latterly worked by steam.

Only the site remains, as the mill was demolished for its bricks and tiles in 1929. The wooden waterwheel survived but an inspection in 1946 stated that it had collapsed.

VALEBRIDGE MILL *Cuckfield*
River Adur – TQ 319 212 – South of Holmbush Farm

This is an ancient site but nothing appears to exist of its early history. An article in the *Sussex Weekly Advertiser* of 9 November 1772 mentions damage to the mill and pond during heavy rain, then occupied by Mr Welfare. Two years later, Welfare was still there, but by 1796, the new owner/occupier was Thomas Ford who remained until 1851. According to *Kelly's* 1862 directory, Peter Holland was the occupier and remained until at least 1907, when all references cease.

An inspection in June 1946 reported that the mill and cottage were derelict but the waterwheel and machinery remained intact.

'Building: Two floors of brick and timber overgrown with ivy and cracked walls. Set into the embankment of millpond. Waterwheel: Iron overshot 10ft in diameter by 5ft 6in wide in an excellent condition. Iron axle shaft. Pentrough: 10in square wooden trough with 6in pipe above. Gate has inscription 'A. Shaw of Lewes 1888'.
Pit Floor: 7ft iron pit wheel, 3ft wallower, 6ft spur wheel and octagonal wooden upright shaft. Stone Floor: 2-pairs of stones (1 peak 1 burr), 4ft 6in crown wheel with pinion drive on south side for sack hoist and winnower. From the west side a shaft worked a wire dresser. Miscellaneous: The west part of the building was the miller's cottage but has been converted to a cow shed. Extensive millpond.'

According to a visit by Frank Gregory in August 1966, the mill had burnt down and then demolished. Today (1995) only the wooden pentrough survives with the millpond idyllic as ever.

WEST MILL *Small Dole*
Tributary to River Adur – TQ 213 138 – At West Mill Farm

A mill was mentioned in 1553 and called 'New Mill', working in conjunction with the nearby

Wood's Mill for many years, but during the 19th century it was used as an oil mill.

The mill was always the junior partner to Wood's Mill as a flour mill but it became a well known oil-seed mill. The sale notice in the *Sussex Advertiser* of 11 May 1847, stated that the demand for oil cake kept the mill in full employment day and night. The mill opened in 1837, run by John Wickens, until it closed and reverted back to flour milling in 1850.

Sale particulars issued by Messrs Challen & Sons advertised both corn mills for sale in July 1853, by order of the Trustees of John Botting, late of Twineham. The 1879 Ordnance Survey map annotates the site as 'woollen' but nothing is known about this venture. *Kelly's* 1882 directory refers to West Mill as part of the farm (not in water) indicating it had obviously closed. The *Victoria County History* states that the buildings were demolished by 1939 although the millpond survived until 1947.

There is nothing to be seen today (1992) and a riding stable now occupy the site.

WOODFIELD FULLING MILL *Cuckfield*
Tributary to River Adur – TQ 314 211 – North of Freek's Lane

The site of the last working fulling mill in Sussex cannot be identified, and the 1873/4 Ordnance Survey 6" map makes no mention of it. According to the 1844 Tithe Apportionment, Thomas Marris was the occupier and Charles Bayntun the owner, but nothing else known apart that it was approximately $1/2$ mile downstream from Valebridge Mill.

WOOD'S MILL *Small Dole*
Tributary to River Adur – TQ 217 317 – At junction of A2037 and Horn Lane

Small Dole is the home of the Sussex Trust for Nature Conservancy with the headquarters at Wood's Mill. The mill, as part of the field centre, is open to the public and is a favourite location for school parties.
This is probably one of the Domesday mill sites, in the ownership of the Bishop of Chichester, while the *Victoria County History* records that the name Wood's Mill was mentioned as early as 1538.

From at least 1828 to 1866, *Pigot's* directories record Jonathan Botting as miller (also for a time at West Mill) after which George Holman took over and remained until 1878. Daniel Atkings, whose stay was brief, followed him. Charles Cook took over and it remained with the family until at least 1927, and it is also recorded that steam power was being used from 1911. In the 1930's, the mill was used as tea rooms with the waterwheel used to power an electric generator. According to an inspection in May 1945, the machinery, except the waterwheel, upright shaft and pit wheel, had been removed and the building used as a workshop with a dynamo powered from the upright shaft. The iron overshot waterwheel 9ft in diameter by 8ft 2in wide, was set below a millpond with an iron penstock inscribed 'Neal & Cooper Millwright 1854'. It is possible that the waterwheel was installed at the same time, but even so, it was completely overhauled in 1895.

After milling ceased in 1927, it was fortunate that the building was never left to deteriorate and Mr J. Douglas Smith bequeathed the mill to the Trust in 1966. In addition to the mill and the 18th century mill house, are 15 acres of marsh, lakes and woodlands and an assortment of outbuildings.

The mill is a fine looking building built in the traditional brick and timber to three floors. As a display feature, a pair of stones were set up on hurstings on the ground floor for demonstration purposes but at the present time the waterwheel is out of use and awaiting repair.

Ashington Mill in the early 1900's (VU)

The rebuilt Ashington Mill in 1939 (AS)

Bolney Mill and pond

Overshot waterwheel at Bolney Mill in 1951

Court Mill in 1992 with its idyllic millpond

Overshot waterwheel at Court Mill in 1992

Sale Notice of Woods Mill (AB)

The ancient Gatewick Mill in 1878

Hammonds Mill in 1911 (VU)

Hooker's Mill prior to its demolition in 1900 (FG)

Highbridge Mill in the late 1890's

The disused waterwheel at Highbridge Mill in October 2000

The disused wheelpit at Manor Mill in 1939 (AS)

Leigh Mill disused in 1953

Manor Mill in 1939 in a ruinous condition (AS)

Manor Mill at Poynings in the early 1900's (RP)

Ruckford Mill and pond in the early 1900's (RP)

Shermanbury Mill in 1904 (RP)

Wood's Mill after closure in 1929 (AS)

Wood's Mill in use as a workshop in 1936 (AS)

Ancient clasp-arm waterwheel at Slaughter Bridge Mill

Slaughter Bridge in use as petroleum depot in 1956

Trustler's Mill in the early 1900's (RP)

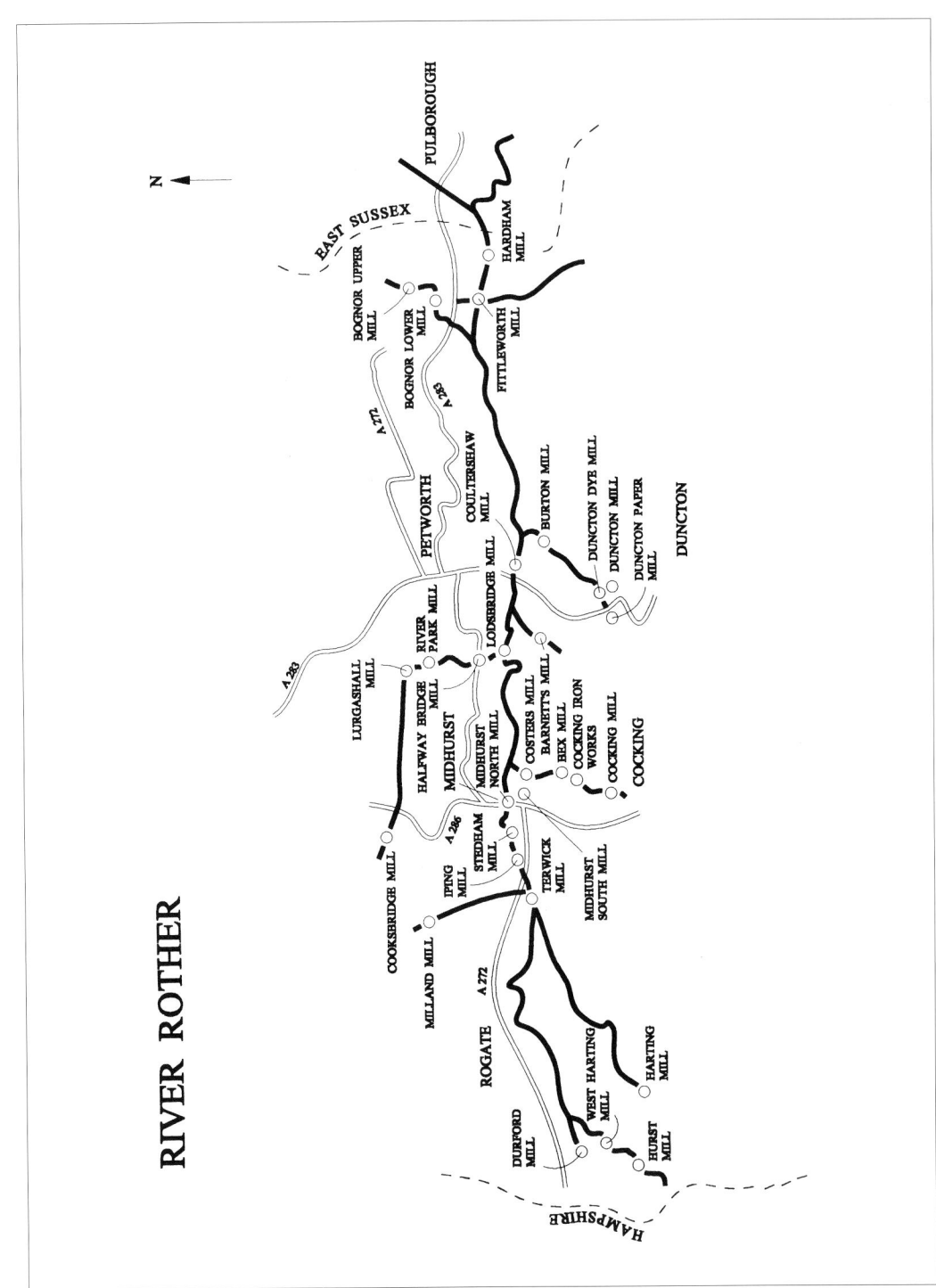

RIVER ROTHER

BARNETT'S MILL *Lodsworth*
Tributary to River Rother – SU 945 195 – 1 mile southeast of Selham

This mill stood in an isolated location between Graffham Common and Lavington Common and although marked on Faden's 1795, there are few references to it.

According to the Tithe Apportionment, in the middle of the 19th century, Richard Porter was the occupier, while later between 1882-1900, William Woods was in control according to *Kelly's* directories. It was described as a 'Flour Mill' on the 1874-5 Ordnance Survey Map 25", and 'Corn' on the 1895-6 map (it stopped work in 1900) while later in 1914, the mill is marked 'Disused'. Surprisingly, considering the poor water supply, the mill had two wooden waterwheels.

According to an inspection in 1936, it was a brick and timbered structure with a cement faced cottage on its northern side and derelict. The dam separated the mill from the millpond, and apart from a few brick footings, the disappearance of Barnett's Mill is complete.

BEX MILL *Heyshott*
Tributary to River Rother – SU 805 187 – 2 miles west of village

Bex Mill was converted into a private house in the late 1970's and in the garden, water pours over a static waterwheel. Although not specified as such in the Domesday Survey, this certainly is an ancient site with a reference in about 1240. Miss E. Gardner has admirably chronicled its history in her article that appeared in Vol 14 in *Sussex Notes and Queries* in 1956, which should be consulted for further and more detailed information.

At the turn of the 19th century the owner/occupier was Richard Cobden who sold the mill in 1802 to James Monk and Jabez Shotter, shopkeepers of Midhurst, who agreed to pay off the mortgage of £800 plus a cash sum. They sold it a year later to Elizabeth Smith for £1,680 and according to the Land Tax
Assessment, she owned the mill until 1813 when it was sold to Richard Moores of Terwick. After his death in 1841, his executors sold the property to John Mills who had been the miller at Bex Mill since 1811. Henry Mills took over and continued as miller until he died in 1876, his son Henry taking over, who set about 'modernising' the mill by replacing the wooden machinery with cast iron and fitting elevators throughout the building and later adding a steam engine. Following his death in 1904, his nephew Henry Wells took over, he being followed by his sons Maurice, Henry and Donovan who continued until 1946, when the youngest son sold the mill to Mr J. Ruthven who carried on milling for a short time.

Bex Mill is a two bay mill, extended over the years, and erected just before the turn of the 19th century as a legal document of 1800 refers to "all that watermill, mill house, cottage, stables and garden premises thereon lately erected".

It is a square building built of whitewashed dressed stone and brickwork, under a half hipped roof, set into the side of a steep millpond embankment. It once contained 3-pairs of millstones, one french burr, the others composition, but in its latter working years, (it closed down in the 1960's) it just ground provender feed, as flour had not been produced here since the end of WW1. The steam engine was replaced by an oil engine, taken out and sold just before the last war. The mill, on a tributary of the Western River Rother, stands on the south side of the lane from Heyshott to the main Midhurst to Chichester road. The stream leaves Bex Mill and flows to Coster's Mill before joining up with the River Rother near Cowdray Park.

BOGNOR LOWER MILL *Fittleworth*
Tributary to River Rother – TQ 006 206 – West side of road below Upper Mill

This was the lower of two watermills that stopped in 1895. It was reputed to be an ancient site but it was always a small country mill serving the community with flour ground from its two pairs of stones. The mill worked in conjunction with the nearby Upper Mill, certainly through the 19th century, and both sites were marked 'Flour' on the 1880 Ordnance Survey 25" map.

The sandstone rubble building is situated behind the larger mill cottage, under a tiled roof, and was worked by a 16ft diameter by 3ft wide overshot waterwheel. The water was fed onto the wheel through a cast iron trough, similar in design to that found at the Upper Mill site. The machinery and waterwheel was removed in 1915, after which it became a house with the large bakehouse removed at the same time. There is little to be seen apart from a plaque on the mill cottage, which is inscribed 'Crowsole Mill. A 12th century flour mill. Ceased working in 1898'.

BOGNOR UPPER MILL *Fittleworth*
Tributary to River Rother – TQ 007 208 – Near Bognor Common

Little Bognor is little more than a few houses north west of Fittleworth. Even so, the village supported two watermills both marked as 'Flour' on the 1880 Ordnance Survey 25" map. In 1822 the occupier was John Reeves, according to a sale notice of 27 May, when it was sold in conjunction with the Lower Mill. The 1841 Tithe Apportionment lists George Sharp as the miller and Melville's directory confirms that he stayed until 1851 when E. Stubbington took over for a few years. Alfred Knight took over with William Stenning being the last miller until closure in 1898.
This was the larger of the two mills, pulled down before WW1, while today just the millpond remains, together with a short section of cast iron section protruding out of the pond embankment.

BURTON MILL *Duncton*
Tributary to River Rother – SU 979 180 – South of Petworth

The Wealden Iron Industry, of which the Burton Mill site was a part, flourished in Elizabethans times and a forge was here in 1667.

On Budgen's 1724 map the site is marked 'Engine to raise water' but the present mill dates from 1780, when it was owned by the Biddolph family, who were resident at Burton Park. From 1830, Joseph Welch was the tenant according to the *Sussex Weekly Advertiser*, but his son Richard from about 1862, according to *Kelly's* directory, succeeded him. Richard continued to at least 1882, after which Henry Challen took over, but he was soon replaced by John Slade and it was during his tenancy that the Burton Park estate was put up for sale, according to an advertisement in 1894. The mill was let to Slade together with a farm of 57 acres, which stretched back to the railway line. The mill had 4 floors, two overshot clasp-arm water wheels, 4 pairs of stones, large corn bins with elevator shoots etc, conveniently arranged with entrance from the road on the third floor, in addition to the entrance from the loading yard below.

The mill closed down just before the turn of the century, and was converted to other uses. The mill was fed by two large 30 acre millponds, which originally drove two overshot waterwheels but, in about 1900, the western wheel was replaced by a turbine to generate electricity to the new owners of Burton Park through a 220 volt overhead DC cable carried on posts through the park for a distance of about 2 miles. The sluice gate to the new turbine bears the inscription 'William Dell & Sons Engineers - Millwrights, Mill Furnishers, Mark Lane London'

In 1929 the remaining wheel was also replaced by a turbine and the mill was refitted out for corn grinding and timber sawing, but this had ceased by June 1947. In the 1960's, subsidence problems with the high pond embankment threatened to take the road and the mill with it and West Sussex County Council, as the Highway Authority, carried out the necessary repairs and acquired the mill and the pond.

In 1978 it was let for flour milling, providing the mill was repaired by the tenant and, as the mill was gutted, a pair of french burr stones, gearing and some free standing iron hursting were acquired from a disused mill located near Cardiff. Members of the Sussex Industrial Archaeology Society, together with help from naval working parties, repaired the western turbine. The mill continued working until 1988, after which the tenants left and despite the council advertising for a new miller in 1990, it remains closed.

Burton Mill is a large 4-storey stone building set into a high pond embankment which has recently, been converted into a house (although the machinery remains). There are also self-contained visitor facilities together with a teashop. The casing of the eastern turbine has been removed and lays discarded outside the mill although the western turbine remains in-situ. There is an attractive mill house to the north of the mill and a plentiful and reliable supply of water, and although the encroaching reeds and water lilies are gradually silting up both ponds, this is still an idyllic site.

COCKING MILL *Cocking*
Tributary to River Rother – SU 880 176 – Centre of village

According to an inspection in January 1961, the mill was in a ruinous state, but a house conversion in 1966 preserved the building from a predictable end. There were five watermills at 'Cochinges' in the Domesday Survey, but little else indicates that this is ancient site.

The mill appears to be about two hundred years old with William Matthews the occupier in 1772, he being followed by William and then John Ayling. John Hammond took over in about 1832 and it remained in his family until commercial milling ceased in 1918, although the mill continued working on a very irregular basis until 1939. The machinery, removed in 1951, was iron and arranged in the standard layout with 2-pairs of french burr stones. The iron overshot waterwheel, manufactured at the nearby Cocking Iron Works, 14ft in diameter by 5ft 6in wide, was removed in 1941, although the wooden pentrough survived for some time.

Cocking Mill is a three floor building set into the embankment of a small millpond with direct access into the top floor from the causeway. On the outside flank wall, adjacent to the filled in wheelpit, are the unmistakable scrape marks from a waterwheel that once turned badly out of alignment.

It is an attractive building of stone under a half-hipped tiled roof, and although the millpond has been drained, the mill is set in a delightful location.

COCKING IRON WORKS *Cocking*
Tributary to River Rother – SU 886 185 – South of Bex Lane

These works were known as 'Chorley Iron Foundry' and manufactured agricultural implements, although work was carried out for local watermills, with the waterwheel at Cocking Mill manufactured here.

It stopped in 1884, when Chorley's business ended in Midhurst, while the wooden buildings, together with the large iron waterwheel, remained for some time after, as they are marked on the first edition of the 1/2500 Ordnance Survey 25" map. The site is set amongst a tangled mass of undergrowth with access strictly private.

COOKSBRIDGE MILL *Fernhurst*
Tributary to River Rother – SU 849 273 – West of Fernhurst Road

This site, tucked away in the northern part of West Sussex, appears to have gone unrecorded even though it was close to the Fernhurst to Easebourne road.

There is little to be seen today (1994) apart from a subtle change in vegetation and a raised embankment that could have contained the mill leat. *Kelly's* directories list a succession of millers here, with Charles West the last in 1905, although the 1913 Ordnance Survey 6" map marks the building as 'Corn Mill'. The site lies on the west side of the road opposite the Lurgashall turn off and according to an inspection of the site in 1939, only the lower walls of the mill remained.

COSTER'S MILL *West Lavington*
Tributary to River Rother – SU 896 207 – By track leading of Selham Road

This mill stands in West Lavington beside an unmade track that runs south from the village. This appears to be a fairly ancient site with a reference in a Deed of 1676, but nothing more is known until the publication of directories in the 19th century. According to *Pigot's* 1839 directory, William Monk was the occupier while *Kelly's* directory records that from 1862 until 1878, Edmund Catt ran it. Arthur Bennett took over and continued until the closure just at the turn of the century, although the mill is later marked on the 1913 Ordnance Survey 6" map.

It appears that the mill lay disused until the late 1930's, when some of the drive machinery was transferred to Midhurst North Mill. An inspection of the mill in March 1939 revealed that it was being used as a pig food store but most of the wooden pit machinery was still in-situ. The overshot waterwheel was manufactured locally at the Chorley's Iron Foundry and had a charmed life, for after the mill closed down it was moved to Lurgashall Mill where it continued working for many years. After the mill closed down it was dismantled and re-erected at the Weald and Downland Museum at Singleton, north of Chichester, where the wheel, in its third environment, continues to run the old mill.

Coster's Mill was converted into a house in 1970, but is an attractive building with its dressed stone and brick construction and a tablet on the east wall is inscribed '1835' which is likely the date of a major rebuild here. There are three floors but the machinery has been removed and the plain pitched roof of the former mill has been altered into a tiled half-hipped roof and the addition of several new windows and a brick chimney, altering its external appearance. An embanked leat provided water to the wheel, but this is now overgrown.

COULTERSHAW MILL *Petworth*
River Rother – SU 972 194 - Beside A265 1 mile south of Petworth

This is one of the important sites of West Sussex as it preceded the Domesday Survey. The last mill was architecturally one of the ugliest and most unattractive buildings you could wish to see, especially so bearing in mind its idyllic location. Aesthetically it was a blessing in disguise when it burnt down in the 1970's. For once, the early history of the mill site is well documented in abundance, but it is not possible to include all this information in this report. Briefly, the mill belonged to Shulbrede Priory from 1240 until the Dissolution, after which Henry VIII gave it to Sir William Fitzwilliam, High Admiral of England from which it came under the direct control of the manor of Petworth.

In more recent times William Dale, the miller, was made bankrupt in 1803, according to the *Sussex Weekly Advertiser*. It appeared that William and Robert Dearing took over until 1841, followed by James Boxall. Maurice Ireland followed in 1856 and it remained in the family until 1905, after which James Gwillim Ltd took over, who later on ran the watermills at Midhurst, Ebernoe and Fittleworth. His wife and then their son carried on the business until the mill was gutted by fire in 1923, so ending more than 1000 years of continuos traditional flour milling at this site. It was a large watermill, built entirely in dressed stone under a tiled roof, with ample storage attached to its east wall, which abutted a wooden pump house. It had a large undershot waterwheel, supplemented later by a turbine. The mill was adjacent to the Rother Navigation (which no doubt expanded its trading limits) and from contemporary photographs at the turn of the century, was at least 200 years old. The replacement mill was in complete and total contrast, being flat-roofed, 5-storeys high, constructed in ferro-concrete looking more like a tenement block! Built when the Planning Laws were easier, it seems incomprehensible that such an ugly building was allowed to be built here in rural and open countryside. Power was supplied by two turbines, one of which was sited in the wheelpit.

Coulershaw Mill was situated on a short length of a mill stream that left the Rother Navigation just upstream from Coultershaw Bridge. The navigation bypassed a narrow and winding section of the river which provided access for boat traffic further upstream. A stepped weir ensured that a constant head of water was available for both the navigation and the mill, and the lock immediately upstream of the bridge. A further benefit came in 1800's when the Petworth to Chichester road was diverted by way of Coultershaw Bridge.

Although nothing remains of the mill, the pump house installed by Lord Egremont in 1782 survived and continued working until 1960. This installation comprises a 11ft diameter by 4ft 6in wide breastshot waterwheel, built of cast iron with wooden paddles and sole boards which drove three triple throw pumps. Water was delivered from these pumps via a 3in diameter pipe nearly 12 miles long, to a 23,000 gallon reservoir on Lawn Hill in Petworth Park, and to a large cistern in Grove Street Petworth. The former supplied all the needs of Petworth Hose, while the latter supplied the town. The pump operated until 1960 supplying the garden and stables at Petworth. The pump and machinery have been restored back to working order by members of the Sussex Industrial Archaeology Society, and is open to the public.

DUNCTON MILL *Duncton*
Tributary to River Rother – SU 964 166 – 2 miles southwest from Duncton

The mill is set into a high millpond embankment and built to four floors, with entry at the front by a small flight of steps into the bin floor of the mill.

The first references are to be found in the Petworth Minister's Account between 1347 and 1373, when the tenancy of a new mill is recorded, because the previous mill and mudwork 'had been shattered and swept away by the flooding of the water.'

The existing mill dates to the latter half of the 18th century, according to a sale notice that appeared in the *Sussex Weekly Advertiser* of 4 March 1776. It was described as a 'newly built Water Corn Mill', and 'that it is built upon a very good principle, never wants water in summer, nor it is subject to being flooded in winter.'

As a sign of the general unrest at the time, the following article appeared in the *Sussex Weekly Advertiser* of 16 March 1795 as follows:

'Two anonymous letters, one addressed to Mr Hamman (should be Hammond) miller at Duncton, near Petworth, in the county, and to Mr Dale, miller at Petworth, threatening to pull down their respective mills, and to distribute the corn found therein amongst the poor, unless the price of flour was immediately lowered. For the discovery of the writer or writers a reward of £150 and his Majesty's pardon to any accomplice, making such discovery, has been offered in the *London Gazette*. Mr Hammond was so incensed by this threat that he assigned a further £100 to the reward money, but the outcome was not recorded.'

Until 1876 the mill was still in the occupancy of the Hammond family, but when Henry Hammond died on 21 November, aged 68, the tenancy of the mill was taken over for a very short period by Leonard Eames (who later moved to Hardham Mill near Pulborough), and then by William Drewitt. Frank Turner became the next occupier in 1887 and it stayed in his family until 1939, even though milling had ceased in 1920. Frank Turner, the last miller, recalled in January 1939 that "The original mill was burnt down and rebuilt about 150-200 years ago. The mill is still in use and is used mostly for working the pump for spraying fruit trees. I was using it for that purpose yesterday. The milling business was finished about 1920. There are three pairs of stones in the mill. My father, Frank Turner, came here in 1887 and died in 1936."

The mill stopped working in 1952 and an inspection in 1994 revealed that the mill was still empty apart from the pit floor where large tanks are used for breeding trout, while outside, at the back, there are further tanks used for the same purpose. The iron overshot waterwheel is still in a fairly good condition and a pencilled inscription inside the mill gives the date ie. 'New iron waterwheel and trough 1882'

Duncton Mill ended its working days driving a water pump and oat crusher. The sturdy wooden upright shaft and 10ft diameter iron pitwheel survive on the pit floor but the wooden spur wheel is broken into pieces. No millstones remain on the floor above, but the wooden crown wheel is still in-situ with two rows of wooden teeth. One set drove the sack hoist (existing) while the other drove a silk flour machine and, via a countershaft, other machinery.

Duncton Mill and farm are grouped together in a small valley from which a relatively small group of springs issue, forming a small tributary to the River Rother, while the necessity to preserve water here is illustrated by the large millpond. The brick and stone mill is in a good condition structurally but access is not recommended, not just because it is on private property, but owing to the poor state of the floorboards (especially so on the stone floor!).

Together with the mill house (part Queen Anne but restored) this is a most pleasant group of buildings set in a small valley in open countryside to the south of Petworth.

DUNCTON DYE MILL *Duncton*
Tributary to River Rother – SU 968 165 – North of Duncton Mill

This was a fulling mill in the 18th century, according to a sale notice that appeared in the *Sussex Weekly Advertiser* of the 4 March 1776, working in conjunction with the nearby Duncton Mill.

In the early years of the next century it appeared to have changed to grinding dyes and was named on Greenwood's map as 'Dyehouse'. In 1951 it was reported that the old dye house was pulled down about 15-20 years ago.

There is no sign of the site and the small fast flowing stream now runs beside a house of modern origin.

DUNCTON PAPER MILL *Duncton*
Tributary to River Rother – SU 958 168 – In grounds of Seaford College

This was the site of a paper mill established by Lord Egremont in the latter part of the 18th century with George Edds the master paper maker in 1810, while the Tithe Apportionment of 1839 refers to Elizabeth Chorley. Thereafter, nothing is known about the venture and there is no indication of the site on the 1914 Ordnance Survey 6" map.

The Petworth to Chichester road (A265) originally bisected two ponds here, but only the west pond is left within the grounds of Seaford College.

DURFORD MILL *Rogate*
River Rother – SU 780 233 – 2 miles east of Petersfield

This small brick built mill is hidden away and appears to have been forgotten as a West Sussex mill, as it is away from a public road and just 200 yards east of the county boundary with Hampshire.

There is a brief reference to 'Durford Mills' in 1620, but it is not until 1738 that we learn of the name of Henry Legge, according to a stone tablet on the east flank wall, while another tablet on the mill house bears the inscription 'H.S.B.L 1779'. *Kelly's* 1895 directory records Rowland Hall until at least 1903, after which W. Waters took over until 1922, with Frederick Brown the miller when it closed down in 1927. It was standing disused in 1939, and the machinery removed by 1954, with the mill used as a toy factory. The present occupiers took over in 1991 and completely modernised the building, adding a new facade on the west side of the mill, and converting it into modern office accommodation.

Durford Mill had a sloping roof on its south side almost down to ground level, while a single storey addition provided some cover for the externally mounted breastshot waterwheel, together with a millers convenience. A short leat (now filled in), fed the water to the wheel and later to a turbine, and the old wooden pegged sluice gate survives marooned without any purpose. At the back, the original wall retains its original features.

FITTLEWORTH MILL *Fittleworth*
River Rother – TQ 009 184 – South of village

This is an attractive building of local stone located on the River Rother, at a point where the Rother Navigation forms an island at Fittleworth Bridge.

The existing mill was reputedly built in 1695 on the site of a fulling mill and throughout the 19th century there were a succession of millers here i.e. 1839 - John Tribe, 1855-1878 - William & Jacob Turner, 1878 - James Farrell /Henry Joyes, 1905-1921 - Henry & Hugh Joyes.

There is some confusion as to how many waterwheels the mill possessed, for a short report prepared by Peter Davies in 1946, makes reference that two of the three waterwheels survived, while Mr Hawksley stated that there was no indication of a third waterwheel position. As the mill contained 5-pairs of stones, it is most unlikely that a third wheel was required.

The last miller was John Gwillim, who also controlled watermills at Midhurst, Ebernoe and the large Coultershaw Mill near Petworth. He removed all the old machinery when he took over in 1921 and the mill was belt driven until it stopped working in 1927 although one of the replacement wheels (installed 1924) was used to drive a dynamo. In 1945 the mill was gutted and used as the headquarters of the Military Police and converted into a house in 1980.

The former mill is set in picturesque surroundings with the waste water from the two empty wheelpits rejoining together, before passing under the road bridge.

HALFWAY BRIDGE MILL *Lodsworth*
Tributary to River Rother – SU 931 220 – Behind Bridge Inn

This was one of the few Sussex mills bordering a main road, and took its name from the bridge that lies midway between Midhurst and Petworth, and was not originally a corn mill.

It is not thought to be an ancient site and the first reference appears on Gream's 1795 map. It is not until the advent of Kelly's directories that some of the millers are known. Arthur Blaker was in control of the mill from 1878, first in partnership with William Cooper and then with his own son. The mill had been rebuilt in 1872 following a fire and was converted to a roller mill until it too was later destroyed by fire in 1904. A 'River' turbine manufactured by J.J. Armfield & Co.powered the mill and according to a testimonial, dated Nov 1903, was 'running satisfactorily and giving every satisfaction'.

On the 3 February 1896 the *Miller* published an article describing Mr Blaker's mill in detail. It relates that he was using the three millstones at the nearby Lodsbridge Mill for grist and barley grinding and was using a granary for wheat cleaning. The grain was then carted to Halfway Bridge Mill for grinding. Both water and steam provided the motive power at Halfway Bridge Mill, with the former of the breastshot variety. A semi-portable engine was housed in a building adjacent to the mill and separated from it by a 'stout' wall. Although the 3-pairs of millstones were retained in the mill and could be used, Mr Blaker installed a Tattersall roller mill in 1895, which, by all accounts, was a success. The article states that Mr Blaker's son was in sole charge.

After the fire of 1904, the mill was rebuilt and continued working until 1911, after which all references to the working mill cease. The building was used as a corn store for some years, but by 1958, it was being used as a toy factory until it burnt down in 1965 according to Frank Gregory.

Today (1994) only the base walls of the mill remain, complete with score marks from a very uneven waterwheel, with the wheelpit filled in. A commercial company specialising in the distribution of creative art materials now uses the building.

HARDHAM MILL *Hardham*
River Rother – TQ 034 178 – West of village

Hardham was once an important Roman posting station on Stane Street, but the village has been by-passed by the busy main road to the coast. The mill was detached from the village centre and the site now lies within a large water treatment works operated by Southern Water.

The River Rother was first penned back at Hardham in 1785 to provide a head of water for the Arun Navigation, in conjunction with Hardham Lock, which was later used by the mill. On the 17 April 1827, the Navigation agreed to grant George Sharp of Guildford, a lease for the right to use the waste water running out of the river near Hardham Lock, for the purpose of a watermill which he wished to erect. George Sharp commenced work in 1828 and used the available barge traffic to transport corn and wheat to and from the coast. The tenancy of the mill stayed within the family until 1886, with Walter Sharp the last family miller. At the same time the mill was advertised for sale according to a notice which appeared in *The Miller* on 3 May;

'To be sold by auction at the Swan Hotel, Pulborough on June 3. Hardham Mill and residence, situate in the parish of Hardham, Sussex. A substantial brick built water and steam mill, driving 5-pairs of french stones and one pair of grey stones, dressing machinery, including silk reel, 3 centrifugals, also rolls and purifier, good smutters and wheat cleaning machinery. The 12hp engine was in excellent condition. Attached to the mill is a good granary capable, with the mill itself, of storing 2000 quarters of corn. Bone crushing machinery and chaff cutter driven by water power. The outbuildings consist of stabling for four horses, cart and wagon sheds, a large stout built store, formerly used as an iron foundry, and a timber built and slated carpenters shop. The house is near the mill. The property is leasehold of which 40 years remain unexpired.'

When the Navigation closed on 1 January 1888, the Superintendent was ordered to keep the head of water for the mill, but the company was soon wound up and sold their property, including the mill, to the Stopham estate.

Following the demise of the Navigation, Leonard Eames, the new proprietor, installed a more efficient roller mill, installed by John Smith of Carshalton, although the extant two undershot waterwheels were retained. Eames had come from Duncton Mill in 1887 and remained until the closure in 1928. In 1937 the Catchment Board, at the request of the Stopham estate, removed part of the weir and the mill was subsequently demolished. An enemy bomb destroyed the mill house in 1941 and in 1948 the water board purchased the site.

Today (1994) only some brick footings remain to indicate the site of the mill, which is on private property.

HARTING MILL *Harting*
Tributary to River Rother – SU 790 197 – End of Mill Lane south of village

Little is known about this mill, at the end of Mill Lane, although there are suggestions that this is a very old site, but references are vague and cannot be verified. The mill was certainly in existence in 1731 when John Bartlett of Harting, miller, appeared in the Sussex Marriage Licences. Two years later Bartlett insured the mill for an annual premium of £300, with similar fire insurance policies issued to James Eldridge in 1780 and 1788, and to William Lillywhite in 1794.

It is not until the advent of directories that the millers are known, i.e. 1855 - John Welch (also at the Ship (PH), 1858 - John Hunt, 1866 - George Chitty, 1870 - George Wild (of Wild & Blackman, grocers), 1882 - Henry Johnstone, after which all references cease and according to a previous owner of the Mill House, the mill was pulled down before 1914.

An inspection of the site (1994) reveals that the east and south walls of the mill remained, along with score marks from a very uneven waterwheel. The mill was fed by a stream which issues from the slopes of Harting Hill, before flowing northwards to the River Rother at Trotton and, along with the attractive mill house, is a most idyllic location.

HURST MILL *Nurstead*
Tributary to River Rother – SU 762 210 – At end of track east of B1246

This mill building lies adjacent to the county boundary with Hampshire beside a large millpond at the bottom of a steep hill. This is an established site as the *Victoria County History* records that in 1558 there was a grist mill called 'New Mill' which replaced a fulling mill that had been in the occupation of John Hall, a clothier from Petersfield.

Throughout most of the 19th century, there was a succession of millers here according to Kelly's directories, i.e. 1845-58-George Newman, 1862-66-William Chalcroft, 1870-74 Messrs Smith & Bell & Co, 1878-William Packham. George Ellis took over in 1891 and continued until the late 1920's, when John Witt-Mann took over. Witt-Mann left soon after moving to Midhurst North Mill although he retained offices in Petersfield. Little else is known about how the mill worked or how many stones it contained except that the 14ft iron overshot waterwheel (the wooden axle of which remains) was restored in 1948 and used to generate electricity, sawing and for pumping, with the former use continuing until the late 1950's. The mill contained iron machinery, some of which remains i.e. upright shaft, pit wheel (approximately 10ft in diameter) wallower and crown wheel, but the poor state of the flooring precludes further accurate measurement. There is a concrete base on the pit floor, which shows evidence of engine assistance.

Hurst Mill is a 3-storey building, with a half hipped stone roof, in a poor state of repair (1994) set into a mill dam. Adjacent to the road to the mill is a house called `Westons', which could have been the mill house, which has datestones inscribed 1540 and 1929.

IPING MILL *Iping*
River Rother – SU 853 228 – South of Church

The site at Iping was recorded in the Domesday Survey and later became a fulling mill. By 1665, there was a corn and malt mill under one roof, after which the site was predominately used for papermaking. This was the last working paper mill in Sussex (two working in parallel), according to A H Shorter writing in *Sussex Notes & Queries* in 1951.

The first reference occurred in 1725 according to a fire insurance policy, which also mentions a corn mill on the site. John Biggs had taken over from 1746 and it remained a family business until it was advertised for sale in 1800. The sale notice in the *Sussex Weekly Advertiser* gives an insight into this mill in that there was sufficient water to work 3 breastshot waterwheels, six white vats, presses, frames, fixtures and every apparatus for carrying out the task. It was during Bigg's occupation that a serious fire in June 1758 destroyed most of the buildings, which had thatched roofs. The site was taken over by Messrs Devaynes & Harrison, who remained until 1808, when the partnership was dissolved. Henry Cooke followed them until he was made bankrupt in 1814, and thereafter, there was a succession of short term occupiers ie. 1813 - William Saunders, 1816 - Smith & Warner, 1819 - William Marshall, 1822 - William Wells, 1824 - George Webb, 1825 - Charles Venables. A change in the occupation saw Benjamin Pewtress and James Low remain from 1826 until at least 1842.

In 1851 there were seven beating engines working here, but by 1866 the mill was not working, until

James Allen took over in 1870, followed by Edward & John Warren who remained until 1899, producing mainly white and coloured blotting paper on a single 60 inch machine. Power was provided by both water and steam and by 1917, the site was renamed 'The Iping Paper Mill Ltd' and produced blotting paper.

The paper mills were destroyed by fire in 1925 and never rebuilt and, according to *Kelly's* directory of 1930, the occupier was Robert Evered, who ran a corn mill until WW2 with the site also generating electricity. Using a paper mill turbine as the power source; this continued to at least 1954, according to a site inspection on behalf of SPAB.

A contemporary photograph taken at the turn of the century shows the site with many large buildings with the River Rother widened upstream. All traces of Iping Mill have completely disappeared, and the 'mill house' is modern and no way connected with the former industry. The river is a shadow of its former self and it is difficult to visualise the size and scale of the important industry that once occupied the whole site.

LODSBRIDGE MILL *Selham*
River Rother – SU 933 211 – West side of Selham

This has become a house conversion devoid of its past use. This is not an old site and is first marked on the Gream's 1795 map, although the *Victoria County History* indicates that it could be older.

Kelly's directories record the occupancy of the mill during most of the 19th century i.e. 1845-54 - Thomas Puttock, 1855-74 - Edward Gadd, 1878-90 - John Budd, 1895-1905 - Arthur Blaker (also at Halfway Bridge Mill & Midhusrt North Mill), 1907-38 - James Morley (later & Son). In its latter working days it was engine assisted but it finally stopped working just before the last war.

The stone built mill, adjacent to the River Rother, was powered by two external undershot waterwheels, working in parallel (both removed for scrap during the last war) which were manufactured by Chorley's of Midhurst. One wheel drove a single pair of stones through a one step gear, with the other driving 2-pairs conventionally. A Tattersall roller mill installed to extend the working life of the mill did not prove to be a success.

The mill was derelict in 1936 and converted into a house in 1960, when all the machinery was removed.

LURGASHALL MILL *Lurgashall*
River Rother – SU 940 259 – Adjacent to Mill Farm south of village

The site of Lurgashall Mill is nothing but a grass bank set below the embankment of the former millpond. Photographs of the mill in 1949 indicate that it was of early 18th century construction (later modified in the 19th century) which contained wooden machinery. A SPAB report in 1954, stated that the mill was in a poor condition, while a later inspection in June 1971, revealed that the interior was liable to collapse. In 1973 the mill was 'saved' when it was presented, by the Leconfield Estate, to the Weald and Downland Museum at Singleton near Chichester. In 1977 the mill was painstakingly dismantled brick by brick, together with the machinery, and transferred to the museum site where it was erected three years later as a working exhibit.

An in-depth inspection of the mill by Sydney Simmons in September 1946 gives an insight into the construction and operation of this typical country mill. It was built in brick and stone, with wood mullion windows under a tiled roof, half hipped at its southern end. It lay partly below a pond embankment from which the water powered a 10ft diameter by 4ft 3in wide iron overshot waterwheel (manufactured at the Cocking Iron Works and originally fitted at Coster's Mill at West Lavington). The water was regulated onto the wheel by a shallow trough with the gate operated by a wooden handle, accessible from a window above the wheel. There were two waterwheels in tandem during the 19th century with separate machinery, but by 1946, only one wheel remained, although the axle shaft of the missing wheel was in position. Inside, the iron pitwheel 7ft in diameter meshed with an iron wallower affixed to the wooden upright shaft. Above the wallower was a wooden spur wheel 6ft in diameter, made in four segments, which once drove 2-pairs of stones. Above this floor was 1-pair of 3ft 10" diameter peak millstones (used for grinding animal feed), and a wooden crown wheel 3ft in diameter which once drove an assortment of ancillary machinery by line shafting. An old notice board inside the mill above the door read 'Wheat at 2d per bushel, Barley at 4d per ditto, Oats at 6d per ditto and Mixed Corn at 6d per ditto'. The report also stated that there was a small wooden wheel on the south side of the mill, which drove a dynamo (nothing is known about this).

This is probably an ancient site, being first shown on Budgen's 1724 map, although there is mention of 'Lurgashall millers' William Warner (1698) and John Mills (1713) in the Sussex Marriage Licenses.

From 1845 until 1874 the mill was run by the Cooper family from which George Payne and A. Smeede each took over for a short period. By 1882, James Wrighton operated the mill and stayed until 1895 at least. It was during this period that the mill stopped working commercially and reverted to grinding animal feed mainly for the adjacent Mill Farm. John Anstey was certainly using the mill for this purpose between 1905 and 1938 (and probably later) before the mill closed down in 1950. Instead of bemoaning the fact that this is just one more empty mill site, the preservation and reconstruction of Lurgashall Mill is a breath of inspiration and is seen and appreciated by the many visitors to the popular museum.

MIDHURST NORTH MILL *Easebourne*
River Rother – SU 889 220 – Adjacent to A272 at Easebourne

This was one of two ancient mills associated with Midhurst and Easebourne dating back to at least 1284 when it was worth 40s, and later in 1467 a contract records the rebuilding of the mill. Midhurst North Mill is a substantially built 3-storey stone building, now covered in grime from the adjacent main road, but even so it has not lost its picturesque appearance. Records indicate that the mill was rebuilt in 1806 and during the 19th and 20th centuries, there were a succession of millers here according to local directories, ie. 1792 - John Tipper, 1826-39 - William Tribe, 1851-74 - John Gosden, 1882-95 - Francis Tallant, 1899 A.J. Blaker & Son (also at Halfway Bridge Mill and Lodsbridge Mill), 1903-27 - John Gwillim (also at Coultershaw Mill) 1934-60's - John Witt Mann & Son Ltd. His son Richard Witt-Mann took over using a fleet of nine lorries to deliver flour to the town and out lying villages. The mill continued working during WW2 employing Italian and German prisoners of war, after which the business became affiliated with Bartholomew's of Chichester until it finally closed down in the early 1960's.

This must have been a profitable mill, with its two low breastshot waterwheels, and the mill house in a continuous range, with direct entry into the stone floor. In 1957, the two waterwheels were still in work, producing flour, with electric power used for the grain cleaning machines. In June 1961, the mill only worked during the afternoons, powered by the smaller wooden waterwheel. The mill was converted into a house with all of its machinery removed, with the smaller waterwheel removed to Haxted Mill museum, near Edenbridge, Surrey.

The mill was prone to flooding on a regular basis and the row of three cottages, once situated on the south side of the mill have been demolished. On the opposite side of the road, the River Rother still cascades down the stone overflow channels, its power now redundant after 1000 years of flour milling.

In the summer house, in the back garden of the mill, was found a wooden spur wheel which was proposed to be converted into a coffee table!

MIDHURST SOUTH MILL *Midhurst*
Tributary to River Rother – SU 887 213 – south side of Midhurst

This is an ancient site dating back to the 13th century, valued at 6s 8d, considerably less that the North Mill. A lease, dated 2 September 1765, between Viscount Montague and Thomas Amber, indicated a change in the trade carried out, with leather dressing taking over from fulling. Thomas Amber's grandson, William, was the occupier in the 1830's, but by 1836 he was in the debtor's prison.

The 1842 Tithe Map Apportionment shows a 'Whiting Mill Yard' here owned by Lord Egremont with James Grist as the occupier. Little is recorded about this venture, except that Grist was still here in 1867, but by then it was certainly not water powered.

The 'mill' had ceased working by the turn of the century and been converted into a house, referred to as 'Old Whiting Mill' on a postcard view of 1904. The site was on the east side of the South Pond from which a stream formerly connected to the terminus of the Rother Navigation, a short distance away. A bricked upped arch can still be seen in the pond wall, marked as 'Penstock' on the 1874 Ordnance Survey 25" map.

MILLAND MILL *Milland*
Tributary to River Rother – SU 837 275 – East of Milland Lane

The village of Milland lies close to the county boundary with Hampshire, and a mill was established here before the Domesday Survey.

There would have been a succession of mills on this site and in the 1781 Land tax returns it was recorded as 'Liggates Mill', in the occupation of Mr Coster. The present mill building dates from 1821, and there were a succession of millers here i.e. 1855 - Henry Wakeford, 1858 - Mr Newell 1882 - James Long. Long was an engineer at the Portsmouth dockyard, but after his death in 1890, his sons had no interest in continuing with the mill. It continued working on a very irregular basis until it closed down sometime in the 1920's and in its latter working days used only 1-pair of stones. Towards the end of the 19th century, engine assistance had been introduced.

Milland Mill is a typical country watermill built in brick to three floors, under a tiled roof, set into the embankment of a once large millpond with an overshot waterwheel. It had been converted into a house by 1937, but unfortunately, all the machinery was removed except the overshot waterwheel, which remained in-situ until the outbreak of WW2. The wheel was earmarked for scrap, but after being removed from the axle shaft, it was just left in the wheelpit. It is cast iron, 14ft in diameter by 4ft 6in wide and bears the plaque 'Weyman & Co. - Guildford'. The mill was fed by a small tributary of the River Rother that issues in the grounds of the nearby Milland Place.

RIVER PARK MILL *Lodsworth*
Tributary to River Rother – SU 942 250 - 1 mile north east of village

This is one of the 'lost' watermills of West Sussex that closed down in the early 19th century, leaving little indication of its existence.

The mill was sited adjacent to the eastern embankment of a millpond, just north of River Park Farm, but the 1840 Tithe Apportionment makes no reference to it. According to a Rate Book, it worked in conjunction with Lurgashall Mill.

STEDHAM MILL *Stedham*
River Rother – SU 863 232 – End of track north of village

The existing brick building at Stedham once housed a turbine used to generate electricity to Stedham Hall. It stood on the site of a roller mill that was pulled down before WW2. A mill is first mentioned on Budgen's 1724 map while the first documented reference appears in the 1754 Marriage Licences when Thomas Pratt, the miller, was named. Between 1766 and 1791 it was in the possession of the Eldridge family, with John succeeding his father Thomas, according to fire insurance policies. In 1828, The Times advertised the sale of the mill following the death of the owner John Knight, the tenant being Frank Gardner, according to *Pigot's* directory. However, in July 1843, the *London Gazette* recorded that he was a debtor in the prison at Horsham gaol. William Ayling who, according to *Kelly's* directories, was here to at least 1866 replaced him, and from then on directories refer to a succession of millers i.e. 1870 - Obadiah Waters, 1874-78 - Luther Knight, 1882-90 - William Bridger, 1895 - James Newman.

According to contemporary photographs, the old mill was substantially built of brick under a tiled roof adjacent to the River Rother. *The Miller* of October 1902 recorded the rebuilding of the mill following a fire the previous year, and that a Simons ½ sack roller mill system had been installed. The new mill, constructed of brick, was built in the no nonsense and practical style of the period. William Rogers was the tenant from at least 1915 but previously in 1907, a directory entry classifies the mill as 'steam', although nothing more is known about this venture, with the mill reverting back to 'water' in 1911. Evered & Stratton had taken over in 1922 with Robert Evered the last miller when the mill closed down in 1927.

The mill site is found beside the attractive mill house at the end of a lane leading northwards from Stedham Church, with the power derived from a short leat from the River Rother.

According to Mr Hawksley, there was another mill shown on Greenwood's map that was situated on a small stream south east of Ash House ¾ mile north west of Stedham Mill. According to the 1845 Tithe Apportionment, the mill had gone.

TERWICK MILL *Trotton*
River Rother – SU 831 222 – South of village

Terwick Mill is situated to the south of the village of Trotton and separated from it by the River Rother, but this is an important mill that not only stopped working in 1966, but also consisted of two mills.

The old timber built mill is of some considerable age, while the newer mill, of four storeys and built of brick and stone, dates from 1745, and it is believed that the older mill was the first mill to stand here. The first documented reference to this mill appeared in 1635 when Constance Glenham was the owner, and a fire insurance policy in 1798 gives James Redmond as the owner and Richard Meeres as the tenant. Meeres was still here in 1814 according to an entry in the *London Gazette*, when he appears on a list of creditors for a Portsmouth flour merchant. The 1839 Tithe Apportionment list a Richard

Meeres, possibly his son, and he continued until at least 1858, from which time James Allen took over, until about 1874. *Kelly's* directory lists Edmund Catt from 1878 and he continued until he was declared bankrupt in 1895, according to the *London Gazette*. His son William continued until he was also made bankrupt according to The Miller, in 1898. This was a time when the traditional methods of grinding wheat were under financial threat from the large roller mills, often steam driven. Jessee Ayling took over Terwick Mills in 1897after leaving Ewell Lower Mill, in Surrey in disgust as the owner erected a large steam driven roller mill there. The mill stopped producing flour in 1903 but continued producing animal feed and other similar products until closure. Frank Ayling took over after his father died in 1917 and, with he and his two sons, were the last millers when the mill finally closed down on the 30 September 1966, as it had became economically impractical to carry on.

The two mill buildings are adjoining with two iron breastshot waterwheels between them. The older mill waterwheel is 14ft in diameter by 4ft 6in wide, with the older wheel slightly larger in diameter and width. The old mill drove 1- pair of stones and was still working in 1939, when the iron axle shaft was replaced by a wooden component, by the millwrights Thompsett's of Guildford. There were 3-pairs of stones driven in the newer mill and it appears that both were working until closure.

The old mill was derelict in 1978 while the larger mill was converted into a private dwelling in 1973 with most of the machinery retained, along with 2-pairs of stones.

This is an important mill site in Sussex and although the tail race has silted up, the buildings outwardly retains their connection with its past use. Access to the mill is strictly prohibited and reluctantly it is not possible to give an up to date description of the mill or the surviving machinery.

WEST HARTING MILL *West Harting*
Tributary to River Rother – SU 782 221 – West side of Park Bridge

Nothing remains to indicate the site, apart from three river bridges and part of a dried up mill leat. To the east of the road are man made water channels that formed part of the Downpark Brick & Tile Works, and the mill may have been a pugmill. On the stream that rises at South Harting are three large lakes and it was the middle pond that probably supplied the power to the mill. It was not mentioned on the 1840 Tithe Map, and its history has disappeared into obscurity.

OTHER KNOWN MILL SITES ON THE RIVER ROTHER

Haslingbourne Mill

A corn mill and a fulling mill were recorded as working at Haslingbourne in 1315, but according to a Bailiffs report, in May 1542, on the parcel of waste land where the fulling mill and corn mill once stood, there was now no trace. A replacement fulling mill and corn mill were working in 1583 with Thomas Libbard the leaseholder. After 1664, nothing more is recorded and the position of the mill is unclear. There are suggestions that the mill, by the main river, was destroyed when the Rother navigation was extended to Midhurst in 1794. As Haslingbourne is away from the main river, it is more likely that the site stood on the small stream, which rises just to the north of Petworth, and passes though the centre of this small community.

The ugly ferro concrete replacement mill at Coultershaw in the 1950's

The attractive Coultershaw Mill in 1906

Bex Mill in 1956 prior to its closure

Bognor Lower Mill as a house conversion in 1955

Mill workers at the doorway of Burton Mill in 1879

A rural scene at Duncton Mill in 1936 (VU)

Burton Mill in use as a sawmill in 1949

The attractive setting of Cocking Mill 1908 (VU)

Horse and cart at Cooksbridge Mill in the early 1900's (VU)

Coster's Mill in 1907

Coster's Mill as a house conversion in 1994

Stone floor at Terwick Mill in 1956 (DW)

Barnett's Mill in the 1930's

The two waterwheels at Terwick Mill in 1956 (DW)

Terwick Mill in 1956 (DW)

Duncton Mill disused in 1957

One of the two waterwheels at Fittleworth Mill (RP)

Fittleworth Mill and pond in the early 1900's (RP)

The unusual roof of Durford Mill in 1953

Hardham Mill with its two external waterwheels in 1916 (VU)

The newly built Stedham Mill with the covered walkway in 1906 (VU)

Stedham Mill shortly before it burnt down in 1901

Hurst Mill in 1994

Boating near Iping Mill

Lodsbridge Mill with its two external undershot waterwheels

Lodsbridge Mill derelict in 1949

Lurgashall Mill disused in 1939

Lurgashall Mill at the Singleton Open Air Museum in 1974

Midhurst North Mill and road bridge

Midhurst North Mill as a house conversion in 1994

Milland Mill as a house conversion in 1951

Milland Mill at work in 1900

RIVER OUSE

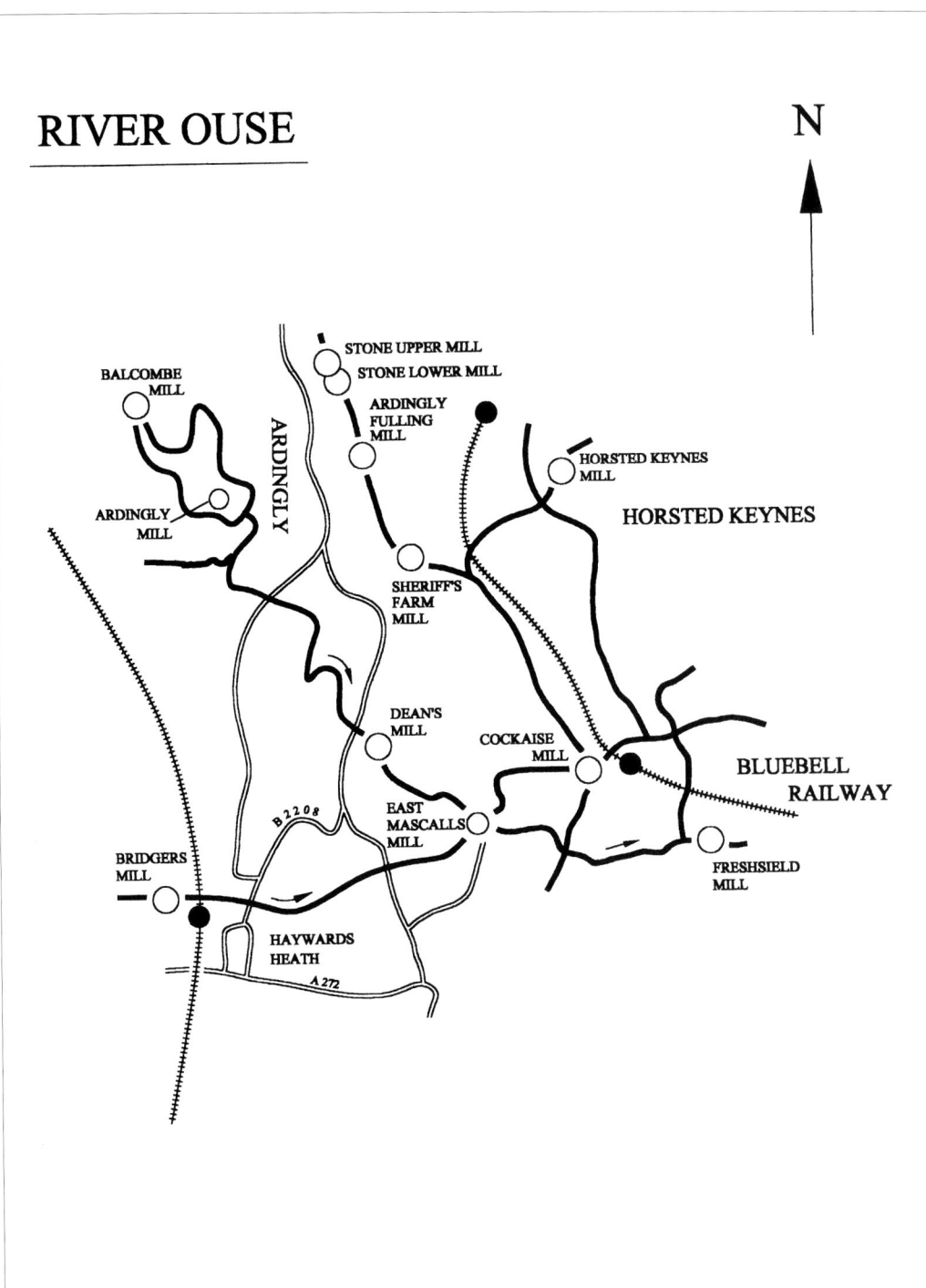

RIVER OUSE

ARDINGLY FULLING MILL *Ardingly*
Tributary to River Ouse – TQ 300 288 – East of Ardingly to Turners Hill Road

The site is easily found, being near to Fulling Mill Farm, but information is scarce. This was one of two fulling mills at Ardingly that both stopped working towards the end of the 18th century. According to the Ardingly Rate Book of 1674, Francis Hamlin was using a fulling mill here. Later, in the *Sussex Weekly Advertiser* of 29 November 1784, the fulling mill, house and farm were advertised as being available for rent, following the death of the tenant Jeremiah Gower; new tenants were not found. Nothing more is known about the mill and there is little evidence to be seen today.

ARDINGLY FULLING MILL *Ardingly*
Tributary to River Ouse – TQ 330 290 – Site under Ardingly Reservoir

There is no opportunity to investigate this former fulling mill site as it is submerged under Ardingly Reservoir. This was the site of an ironworks as the 1674 Ardingly Rate Book refers to John Spence as the tenant of the 'hammer' working here. The fulling mill developed after the demise of the iron industry and worked until the late 18th century.

Both industries are referred to on the 1879 Ordnance Survey 6" map as well as Fullingmill Wood and also East Hammer Wood, both of which succumbed, to the reservoir.

BALCOMBE MILL *Balcombe*
Tributary to River Ouse – TQ 318 306 – South side of road east of village

This was a 3-storey mill standing on an ancient site at the bottom of a steep lane east of the village, converted into a house, but now only 2-storeys high. It stopped working in the 1930's and remained empty and derelict until quite recently. The brick built lower pit floor, including the wheelpit, was filled in when it became a house and now looks rather odd with weatherboarding down to the ground floor.

The mill stands on the edge of Ardingly Reservoir, built between 1975-78, and originally Southern Water Authority wanted to demolish the mill and landscape the site, but eventually a compromise was reached the result of which can be seen today.

An inspection of the mill in 1948 gives an insight into its composition when it was lying derelict: 'Wheel now collapsed and machinery out. Shaft 92" square iron. Wheel was overshot about 16ft by 5ft, wood rim arms and sole, iron buckets, naves were small castings keyed close to the shaft. Wheel was enclosed at east end, fed by twin pipes and gate box with tray. Mill two floors of white painted wood over ground floor of brick. Latter below road level. Slate roof. Waterfall and small pond on east side of mill, 4-pairs of stones'.

The main machinery had been taken out by 1957 but in 1976, under the threat of demolition, permission was obtained to remove loose items in the mill and transfer them to Ifield Mill. The items included the sack hoist, grain shutes, beam scales and the remains of a wire machine. It was also hoped at the time to remove the wooden upright shaft, but this had dropped down in the pit floor at such an angle it could not be removed.

This was a corn mill and had no connection with the iron industry. In the 1803 Defence Schedules, John Booker could supply four sacks of flour daily if wheat was supplied, and according to the 1839 Tithe Apportionment, he was still there. Directory entries give George and James Booker from 1845 until it was taken over by the Turner family. At some later date the mill was run by Jenner & Higgs, based at Bridger's Mill at Haywards Heath.

The mill was fed by a large millpond, now a different shape from that 100 years ago, when it was L-shaped and elongated, and extended westwards right up to the delightful mill house, parts of which date from the 17th century.

BRIDGER'S MILL *Haywards Heath*
Tributary to River Ouse – TQ 330250 – West side of Balcombe Road

The town of Haywards Heath developed and expanded following the arrival of the Brighton railway in 1841. The 1874 Ordnance Survey 6" map shows the mill, with two large millponds, in an isolated position north of the railway station, far different from today, where large housing estates surround the site.

A small tributary to the River Ouse provided the power to this large watermill, whose outstanding feature was its 26ft 3in diameter waterwheel, but little can be found today. This appears to be a fairly ancient site as there is a reference in 1629 when the mill was mentioned during the Beating of the Bounds, but there are few references until the end of the 18th century. In July 1800, Bridger's Farm and Bridger's Mill, together with a piece of land and a windmill, were offered for sale in the *Sussex Weekly Advertiser* following the death of Thomas Kennard and it appears that the property was purchased by John Pace, for the Cuckfield Parish Rate Book for 1801 states 'Two mills and farm, Mr John Pace, £34'. In 1805, the windmill was offered for sale but was demolished or moved soon after, but Pace continued in occupation until he offered the mill for sale in The *Sussex Weekly Advertiser* of 20 March 1820; the property being purchased by John Turner and, for some time afterwards, the mill was recorded as 'Turners Mill.' According to the 1843 Tithe Apportionment, Turner was still in occupation but in 1854 the *Sussex Advertiser* reported the sale of the mill following his death: 'To be sold by auction on the 28th June next, by order of the executors of the late Mr John Turner.

Lot 1. comprises a substantial brick built, tile-healed Water and Steam Corn Mill known as the Bridger Mill, driving four pair of wheat and one pair of oat stones, with a 26 feet 3 inches diameter over shot iron wheel and conducting trough; an engine, with covered boiler, complete (by Hague of London) and fitted up on the most approved principle by Mr Haw, and under the kind inspection of Mr Craven, superintendent of the engineering department of the London, Brighton and South Coast Railway Company, situated in the parish of Cuckfield and within a quarter of a mile of Haywards Heath Station. To view apply to Mr E. Turner upon the premises.'

Ellis Turner continued as miller but by 1857 William Jenner and his son had taken over and in 1861, George Bailey joined them in a partnership, but he left in 1866. William Jenner sold the mill to his nephew Samuel in 1877 who later went into a partnership with a young man called Caleb Higgs. In 1880 the partnership left to take over the nearby Dean's Mill at Lindfield, but the rebuilding of this new mill resulted in the venture failing, and they had left by 1903.

Bridger's Mill was sold for £1,500 in 1880 and the mill continued to trade under the name of Jenner & Higgs until taken over by G.W. Bailey & Son in 1887.

The large waterwheel powered 5-pairs of stones but after a fire partly destroyed the mill in 1915, steam, gas and electricity replaced it. The Hornsby gas engine had its own producer plant until it was converted to town gas in 1921. The mill lay on a downward slope with a banked millpond but it is difficult to comprehend that there was a sufficient head of water to power a 26 feet overshot wheel. (The waterwheel and pentrough were manufactured at the Regent Iron Works in Brighton). Apparently, water was conveyed to the top of the wheel by a large iron trough, but despite the small size of the millpond, the power of this arrangement must have been considerable.

The mill was erected in 1840, and rebuilt following a fire in July 1915 and photographs shows two large 4-storey buildings together with a low 2-storey addition on its eastern side. The mill stopped using water power in 1918 but it carried on a casual basis until closure during WW2. The millpond was drained during the winter of 1964, dug out in the following spring and laid out for residential roads in the summer. The present road cuts right across the former pond from its NW corner to the SE corner. The notice to demolish the mill buildings appeared in the *Evening Argus* of the 4th October 1968 and it was demolished the following year. A large housing estate covers the site of the old millpond and little remains apart from a 3ft diameter culvert, which carried water from the wheelpit.

COCKHAISE MILL *Lindfield*
Tributary to River Ouse – TQ 377 258 – North side of Monteswood Lane

Cockhaise Mill has been converted into a house but retains its original shape and complements the nearby 17th century Cockhaise Mill Farm. This was a typical rural Sussex watermill with brick to the first floor with weatherboarding above, situated west of Freshfield Halt on the Bluebell Railway. The building dates from the middle of the 18th century and could quite possibly be the only mill to stand on this site.

The 1803 Defence Schedules give Anthony Harland as the miller but by 1808 Walter Hurst had taken over and it remained with the family until at least 1855. After a brief stay by Luke Godley, John Comber took over the mill and it remained in his family until closure in about 1936, with William Comber the last miller. There was no engine assistance here and its late closure bears testimony to the large millpond and also to the work generated from Cockhaise Mill Farm, as it ended its days providing animal feed.

On the north side of the mill, the framework of an iron overshot waterwheel, 12ft in diameter by 5ft 6in wide exists, bearing the inscription 'A. Shaw Lewes 1883'. According to an inspection of the mill in June 1946, the waterwheel was in a bad condition and only the pit wheel, wooden upright shaft, some millstones (it originally had 4-pairs) and a bolter remained. The mill remained derelict for many years until it was converted into a house and advertised for sale in February 1966, but overall this is a tasteful restoration although, quite naturally, extra windows had to be installed.

Cockhaise Mill is situated on a small tributary to the River Ouse with the water level of the millpond level with the first floor of the mill, but after the mill stopped working the pond was filled in. As the mill is near to the River Ouse, both the mill and the farm have suffered from flooding, as they are both on ground which is scarcely higher than the main river.

In addition to the main mill, the 1879 Ordnance Survey 6" map shows a 'Corn Mill' to the north east on the west bank of the first bend of the river above Wildboar Bridge. A similar map in 1912 marks this building as 'Disused' but it had been dismantled by 1930. This obviously worked in parallel and in conjunction with the main mill, and therefore is not identified as a separate site.

DEAN'S MILL *Lindfield*
River Ouse – TQ 354 261 – At end of private road east of village

In early days, the parish of Lindfield was granted to the college of South Malling at Lewes with the Dean, Vassal to the Archbishop of Canterbury. The Dean, amongst other tithes, received money from the mill and it was from this association that the mill was named.

It is thought that there was a mill here in 761, but one was certainly grinding during the Domesday Survey, and throughout most of the next 900 years. The present mill dates from 1880, while the previous mill built in 1761, was a paper mill in the possession of the Pim family from 1773, with Francis Pim the occupier. James Pim took over soon after and built a new paper mill at Sharp's Bridge, near Piltdown in East Sussex, which he ran in conjunction. There was a combined paper and corn mill operating at Lindfield and both trades continued to at least 1850, when paper making ceased. Pim had left by 1858, thus ending a long and well established family business here. Robert Jenner was the next occupier of the flour mill with his son, Samuel, taking over by 1874, and the flour mill worked in conjunction for some time with Bridger's Mill at Haywards Heath. Samuel Jenner and Caleb Higgs moved to Dean's Mill but, in 1880, a new mill had to be erected after a storm had totally destroyed the old mill, however the business venture failed and by 1903 they were no longer there.

Kelly's directories record that between 1903-1905 Thomas Rose was the mill manager but by 1907, Alfred Shepherd had taken over the managers post and continued here until 1934 at least. Previously in October 1926, the mill had been offered for sale and the auctioneers survey described the mill as 'built of brick and timber, 4 floors. Sack hoist driven by waterwheel. Four pairs of stones and various machines. On River Ouse. 14ft iron breastshot waterwheel. Ground floor granary and office. Bins above, one of which held 100 sacks. Machinery includes a chaff-cutter, flour screen, oat crusher. Smutter and separator, bran sifter, cake cutter, and winnowing machine. Grind-stone and three loose millstones. Good supply of water in dry summer. Average net profit of working during the past 6 years being £275 per annum. Large scales, old and well made.'

In 1935, Mr & Mrs Horsfield bought the mill and continued to produce stone ground flour until they retired in 1957. The mill carried on working under new ownership until 1976, latterly producing flour marketed under the name of the 'London Health Centre Ltd', with Bernard Lywood the last miller, after which all milling ceased.

The mill is in a workable condition, although the iron breastshot waterwheel has lost most of its buckets. The wheel measures 17ft 6in in diameter by 4ft 6in wide of the compass arm type on an iron axle shaft, with a rack and pinion sluice gate. If the waterwheel was repaired there is no doubt that the mill could easily work again, as the present owners have kept the mill in an excellent condition. Some of the electrically powered machinery is fairly modern, but even this does not detract from what is a splendid and rare example of a complete rural watermill.

The drive and associated machinery is set out in the standard layout as follows:

PIT FLOOR: 10ft diameter all iron cast iron pit wheel, 3ft 3in diameter wallower attached to cast iron upright shaft that turned a 7ft 8n diameter iron spur wheel and four millstone pinions. There is an electrically powered weighing and bagging machine manufactured by Southall & Smith Ltd, together with a set of old beam scales. Three flour shutes are still in place.

STONE FLOOR: 4-pairs of stones complete with wooden tuns, horses and hoppers, together with another pair standing on their ends. There is a wooden rack containing 3 iron tramstaffs and an electricity transformer panel. A door leads off this floor, via a walkway, to the adjacent store building.

An iron 5ft diameter cast iron crown wheel with upward facing teeth drove a combined smutter and separator on the floor above, while another shaft drove the sack hoist which has an iron pulley and idler pulley to tighten the belt. Other counter shafts worked a vibrator feed for the flour machine and combined elevator and worm screws.

AUXILIARY FLOOR: 29ft long flour bolter, without its cloth, combined smutter and separator driven from the crown wheel below and a free standing oat crusher. There is a governor for all the stones together with an assortment of wedges and mill bills. There were handles to operate the sluice gates from every floor of the mill.

BIN FLOOR: meal bins still intact and boarded over.

The mill stands adjacent to the south bank of what was once the original course of the River Ouse, but following the opening of the Ouse Navigation (in 1809) the new line of the river by-passed the mill slightly to the north, which necessitated the construction of a small island to house the northern bearing block of the waterwheel. After passing the mill, the water rejoins the Navigation just below what the 1875 Ordnance Survey 25" map mistakenly names as 'Pin's Lock'.

This is an attractive brick and weatherboarded mill set in a quiet and idyllic location at the end of a road that is strictly private with the mill house opposite of an earlier date. The mill can be seen from the towpath on the opposite side of the river.

EAST MASCALLS MILL *Lindfield*
River Ouse – TQ 365 254 – 1 mile east of the village

The last mill on this site was a small wooden structure located beside the River Ouse. The mill was mentioned on Budgen's 1724 map and marked 'Disused' on the 1874 Ordnance Survey 25" map, and it must have suffered from the competition from the nearby Dean's Mill just upstream. In the 1803 Defence Schedules, John Stevens was the miller but thereafter, the mill remained in the tenancy of the Ansell family until closure. As the mill site is adjacent to the River Ouse it was prone to flooding, and on the 13 November 1852 the *Brighton Herald* reported such an occurrence:

'The late floods. We regret to state that the damage sustained by Mr Ansell, at East Mascalls Mill, in this parish (Lindfield) by the late floods, will amount to some £60, and in addition to this, we are sorry to record that the bridge which crosses the stream running from the water wheel has been completely destroyed by the penetration of the powerful current affecting the foundation and the public highway lane is rendered impassable. Mr Ansell refuses to repair the latter.' Another report in the *Sussex Advertiser* reported that flooding was so bad that the riverbank was washed away. It is most probable that the mill never recovered from this disaster but the thatched mill house remained until the 1920's.

Nothing remains of East Mascalls Mill and the site forms part of a field close to the river, while on the east side of the road is a disused lock, a relic from the Ouse Navigation.

FRESHFIELD MILL *Horsted Keynes*
River Ouse – TQ 386 245 – Northeast of the Sloop Inn

This was a tall mill of brick and weatherboarding, of similar shape to that of the nearby Dean's Mill. It stood in open meadowland just north of the Ouse Navigation (which it obviously used for trading) and was near the site of Freshfield Forge, which was established before 1574 and in ruin by 1664.

The mill was in the occupation of William Arnold in 1826, but Edward Stephens had taken over by 1839. *Kelly's* directories record that Henry Gasston controlled the mill from 1866 until 1911 and it was during this time that the 1874 Ordnance Survey 6" map marks the site as 'Corn', indicating its use at the time. After a short occupation by Reeve Bros, the last recorded miller was J.Luckens.

The last directory entry is 1918 and according to Frank Gregory it was derelict in February 1936. The mill was powered by a large external iron undershot wheel, 16ft in diameter by 6ft 4in wide, which carried the inscription 'Laurence Maresfield', and in 1946, and this was dug out and exposed for an unknown reason (probably for scrap). Water was fed from a leat taken off the navigation from which it passed under the road onto the site. This leat is now (1994) just a depression in the ground, but the sluice gates remain.

The mill was demolished in the 1950's, but now only its brick base remains, with a flat roof, and is used by the farm. The nearby farmhouse dates from 1550, when the pollution and noise from the forge must have been unpleasant, far different from the peace and quiet today.

HORSTED KEYNES MILL *Horsted Keynes*
Tributary to River Ouse – TQ 380 287 – East of Mill Lane

This is reputedly one of the oldest watermill sites in Sussex, predating the Domesday Survey.

One of the first later references appeared in the *Sussex Advertiser* of February 1783 when the miller, Abraham Jackson, was mentioned. George Ross was the occupier from at least 1828, he being succeeded by John Coomber, according to the 1839 Tithe Apportionment. The directories give both Coomber and Edward Stevens at the mill, both described as 'farmer and miller', but by 1861 Mrs Coomber, the proprietor, was reported to be quitting the mill, according to a report in the *Sussex Advertiser*. George Friend took over and worked until he died in 1889 aged 70; his eldest son succeeding him to at least 1914, after which Robert Chalmers took control at a time when only animal food was being produced. All directory entries cease after 1934, but the mill continued working occasionally until its final closure in 1948.

Horsted Keynes Mill is a 3-storey building with brick to the first floor and tarred weatherboarding above, and would appear to date back to the middle of the 18th century. It contains 2-pairs of stones (1 Burr, 1 Peak) and the machinery is a mixture of wood and cast iron, with a 6ft diameter wooden spur wheel. The cast iron 8ft diameter pit wheel, wallower and upright shaft only date from the turn of the last century. The overshot wooden waterwheel, 18ft in diameter by 3ft 4in wide, is set on a wooden axle shaft with a shallow wooden trough carrying the water from the millpond into a 14" pipe onto the wheel, which turns on a frequent basis. Water would not seem to be a problem here as there are a series of ponds upstream of the mill, but even so, there is an external pulley wheel at the front of the mill which would suggest that auxiliary tractor power was used.

The owner must be congratulated on his time and effort in restoring, and preserving this mill and, with the timber framed mill house, it forms a most picturesque setting, tucked away at the end of Mill Lane.

SHERIFF'S FARM MILL *West Hoathly*
Tributary to River Ouse – TQ 360 288 – Opposite Sheriff's Farm.

This mill was demolished in 1931 shortly after it had stopped working. Little documentary information exists about the mill, but a contemporary photograph of the mill at the turn of the last century indicates a large milling concern. The mill was constructed of brick and timber with mansard roofs and constructed at right angles to each other. There was a large boiler house extension on its southern side surmounted by a large chimney. The waterwheel was enclosed and records state that this was 9ft in diameter by 6 ft wide.

The mill first appeared in the 1803 Defence Schedules when the miller, Walter Hurst, could supply a sack (280lbs) of flour every 24 hours, if the wheat supplied. Later, according to the 1841 Tithe Apportionment, John Arnold was the occupier and he ran a nearby windmill in conjunction until at least 1882, according to *Kelly's* directories. From then on, until closure, Arnold Friend ran the mill as part of the Brook House estate.

Only the remains of the wheelpit and the outline of the millpond survive to indicate the site.

SLAUGHAM MILL *Slaugham*
River Ouse – TQ 258 277 – On footpath south of village

The large millpond at Slaugham was established for the iron industry, with a furnace working here in 1574. After the cessation of the industry in 1653, a corn mill was established at the eastern end of the pond. The last mill replaced an earlier structure destroyed by fire in January 1795.

The occupier of the new mill was William Heaver and in the 1803 Defence Schedules, he could supply six sacks of flour in 24 hours, if the wheat was supplied. In September 1824, the *Sussex Advertiser* reported on the dissolution of the partnership between Heaver and Cook who were running both the watermill and the windmill at Handcross. It was resolved that Heaver should run the watermill and Cook the windmill.

According to *Kelly's* directory, John Agate was the miller in 1851, while from 1858, Henry Marchant took over, with William Taylor the last miller. The *Sussex Advertiser* of April 1873 recorded the sale of implements from the mill, owing to the liquidation and closing down of the business. After the closure, the internal overshot waterwheel was removed and a turbine installed to generate electricity to Slaugham Manor until 1900, after which, it became a store.

Unusually, the mill has survived the years, although derelict, and it is believed that the turbine is in-situ, the machinery having being removed in 1937. It is a three storey building of brick, timber and stone under a tiled mansard roof. The mill is adjacent to a footpath running southwards from the Parish Church, with a direct entry into the stone floor.

STONE LOWER MILL *West Hoathly*
Tributary to River Ouse – TQ 346 317 – East of Ardingly to Turners Hill Road

The sites of two largely forgotten watermills lie in a shallow valley, near to Wakehurst Place. Stone Lower Mill was the larger of the two mills, both of which were demolished at the end of the C19th. There were two ponds here with the Lower Mill situated below the southern pond embankment. The mill appeared on Budgen's 1724 map, but there was a reference to the mill in Vol 48 of the *Sussex Archaeological Collections*, where details relate to the mill in 1680. The *Sussex Weekly Advertiser* of August 30 1773 advertises the sale of the mill by William Shaw, both as owner and occupier. The next reference appears in the 1803 Defence Schedules when the miller, John Hollands, could supply two sacks of flour every 24 hours if the wheat was supplied. The mill remained within the family until at least 1882, with

Richard and then Thomas in control.

The 1879 Ordnance Survey 6" map lists the mill as 'Flour' which indicates that it was producing flour when most were just grinding animal feed. However, according to a later edition, published in 1895-6, the mill was marked 'Disused'. A postcard view of the mill taken in the early years of the C19th century, shows it was typically constructed in timber and tile with a large overshot clasp arm wooden waterwheel, while the extension to the mill, on its southern side, was supported on timber piles. Of the mill, nothing remains apart from the millpond and a wheelpit, but yet another mill has disappeared into obscurity.

STONE UPPER MILL *West Hoathly*
Tributary to River Ouse – TQ 346 318 - East of Ardingly to Turners Hill Road

This was the smaller of the two watermills situated on the southern side of the upper millpond and little information survives. According to *Pigot's* directories, Richard Streeter was the miller here from 1832 until 1842, and it no doubt worked in conjunction with the Lower Mill. It was listed as 'Corn' on the 1879 Ordnance Survey 6" map but a later edition in 1895, marks the mill 'Disused'. The site of the mill is on strictly private land hidden in dense woodland.

A winter scene at Dean's Mill in 1958

Balcombe Mill and pond in 1913 (PA)

A peaceful scene at Balcombe Mill in the early 1900's (VU)

House conversion at Cockaise Mill in 1955 (AS)

Balcombe Mill as a house conversion in Sept 2000

The attractive Horsted Keynes Mill in 1932 (AS)

Dean's Mill in 1994

Dean's Mill in the late 1890's (VU)

The stone floor at Dean's Mill in 1994

Bridger's Mill and pond in the early 1900's

The extensive Sheriff's Farm Mill (VU)

Freshfield Mill with its large watermill in 1938 (AS)

Stone Lower Mill at Ardingly in the early 1890's (PA)

Slaugham Mill with its mansard roof (FG)

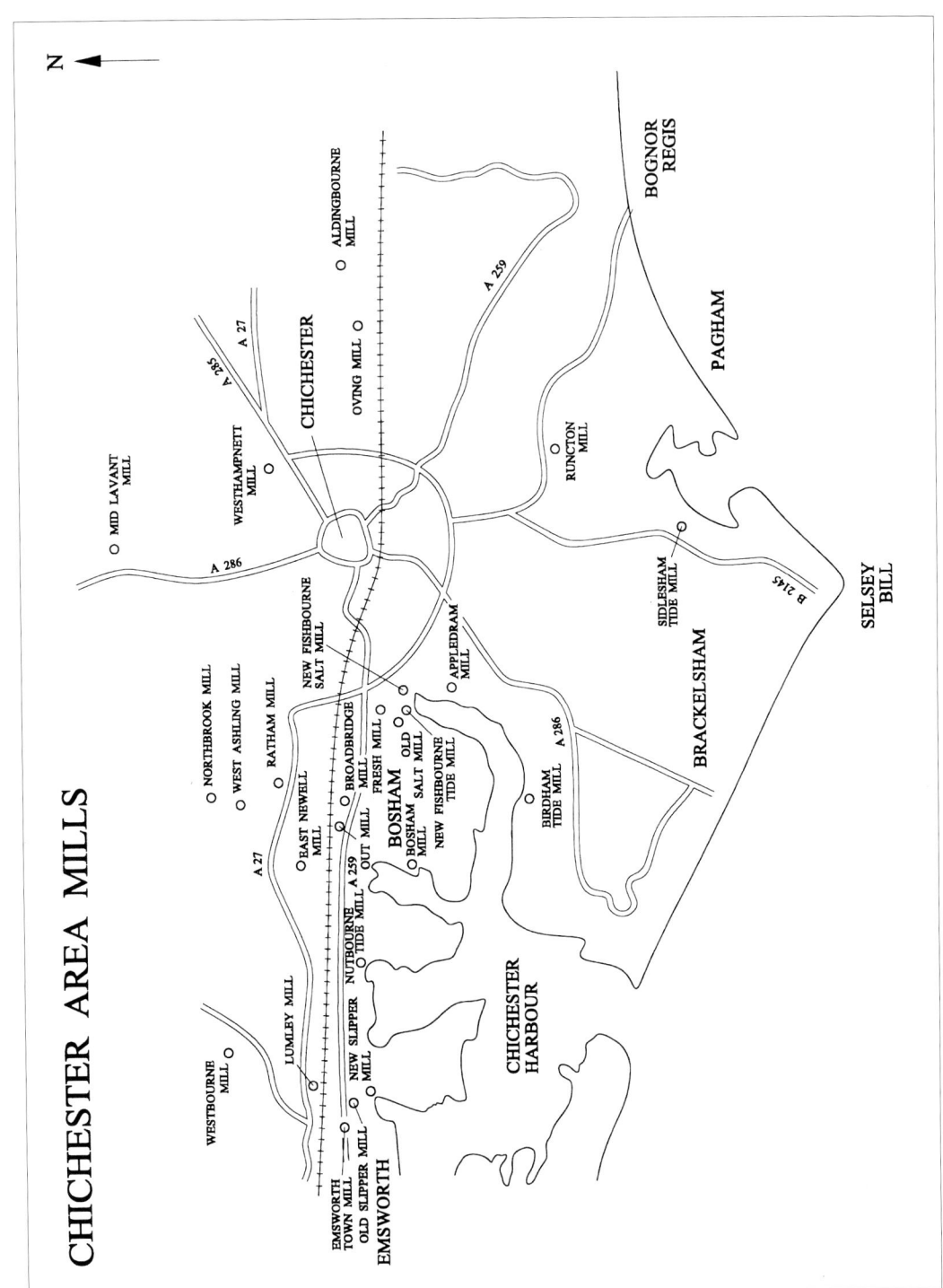

CHICHESTER AREA MILLS

ALDINGBOURNE MILL *Aldingbourne*
Aldingbourne Rife – SU 925 049 – At the end of private access track

Aldingbourne Mill, of 19th century origin, was the last of several watermills to occupy this site. A watermill is mentioned in 1535 while later in 1631, a mill was standing at the 'head of a great pond' leased to Maurice Scull.

The mill worked in conjunction with the local windmill, according to an extract taken from the sale notice that appeared in the *Sussex Weekly Advertiser* in March 1836:

"To be sold. Aldingbourne Wind and Water Mills. An Overshot watermill, built with flint and stone, containing three floors, two pairs of stones and a dressing machine, to which are attached a cottage, stables, piggery, cow-house, a pond about two acres, and nearly the same quantity of dry meadow. These premises are situated in a good corn country and populous neighbourhood. Also there was a windmill, lying in the same parish, on the north side of the Arundel road. Both mills are in the occupation of Charles Smith, who will give up possession on the 30th March".

The new owners or occupiers are not known, but according to *Kelly's* 1845 directory, Richard Boniface was the occupier from at least 1851 until 1899, with both the water and windmill under the control of the Baker family, but shortly before they left, the windmill stopped working. A Mr Chatfield took over and worked the watermill for a few years, but William Hemsley soon followed. It was Hemsley who was responsible for introducing a steam engine and, to accommodate the engine and boiler house, a lean-to was added to the side of the mill. The mill closed in 1914 with the adjoining mill house restored into a house in the 1950's, the mill and most of its machinery fortunately retained.

Aldingbourne Mill was a fairly large concern with 3-pairs of millstones powered by an external iron overshot waterwheel, 9ft in diameter by 7ft wide, manufactured by J. Chorley of Midhurst, which is now little more than a framework. Attempts to remove it for scrap were abandoned due to the problems with access. Inside, the iron pitwheel, 8ft in diameter, survives along with the underdriven millstones, which are set up on a low wooden platform on the pit floor.

The ivy covered mill is used for storage, but is kept in a good state of repair, and is approached by a narrow private road that runs alongside the length of the millpond.

APPLEDRAM SALT MILL *Appledram*
River Lavant – SU 842 038 – West of Appledram Lane

This was never a tide mill as it was worked by the River Lavant. It was situated west of Appledram Lane, where two streams of the River Lavant cross the road. Although marked on the 1813 Ordnance Survey 1" map, a sewage treatment works has long since obliterated it.

BIRDHAM TIDE MILL *Birdham*
Chichester Channel – SU 825 013 – Northwest of Village

Of the seven tide mills that existed in West Sussex after 1724, only Birdham has survived, albeit now used for a difference purpose. Tide mills were often established in remote locations and this was certainly true at Birdham, which was no more than a small group of cottages, a church and an entrance to a canal. The mill building was saved from demolition when it was purchased by a local yacht club in the 1930's.

The original timber mill, rebuilt in 1768, had a brick storage extension added in 1891 and contained 3-pairs of french stones, and 2-pairs of peak stones, with the power provided by two waterwheels, one of which was inside the mill. Tide mills need a large water retention area and at Birdham, there were two large ponds with a combined area of 30 acres, which gave a 12ft head and provided a working period of six hours.

The ownership and occupation of the mill were under the control of the Farne family for many years but it stopped working in 1935, and by 1936 had been gutted of its machinery, apart from the upright shaft and waterwheels (now removed). After the closure, the large millpond provided an ideal location for a yachting marina, with lock gates being installed in 1939 to regulate the water level in the basin. Access to the mill is through a busy and private boat yard.

BOSHAM MILL *Bosham*
North Brook – SU 804 038 – Adjacent to Bosham Harbour

Used extensively by settlements over the years, this is probably an ancient site, in existence before the Domesday Survey, with the present mill building dating from the 18th century. It is constructed in brick and timber under a tiled roof, and, from its exposed position overlooking Bosham Harbour it must have been in need of constant maintenance.

It is located at the end of a long leat that runs through the centre of the village, from where it passes under the mill building. The width of this leat, just upstream of the mill, was once wider but extensive regrading and landscaping have reduced it to a narrow channel.

The first documented reference appeared in a Sun Fire Insurance Policy dated 8 March 1715, when John Williams, the miller, insured his goods and commodities held in the mill, this being one of the earliest insurance policies for a watermill. The development of the present group of mill buildings is difficult to determine with the main body of the mill, running north to south, appearing to be the original mill dating from the 18th century. The extension at right angles was first shown on the 1875 Ordnance Survey Map 25".

Just before the turn of the century, two new overshot waterwheels in tandem were installed which replaced the previous single wheel. Owing to the limited water supply, a rather complicated method of channelling water to the two waterwheels was employed to ensure that they could work at the same time, if required. These wheels are not enclosed, although due the slope of the building they appear to be so. Both waterwheels are constructed in iron with the 'north' wheel 10ft in diameter by 6ft wide, and the 'south' wheel 9ft in diameter by 6ft wide, (both with 24 iron buckets) and although both are of different diameters, the tops of both wheels are level. It is thought that they drove 5-pairs of stones between them.

According to directories, the mill was in the occupation of Thomas Boone & Sons from 1845 to at least 1859, after which they were replaced by Thomas and Henry Young while later, the mill continued in the ownership of the Brown family until references cease in 1934.

Bosham Mill is used as headquarters for the local yacht club, and apart from the addition of several extra windows, has retained its attractive charm and appearance.

BROADBRIDGE MILL *Bosham*
North Brook – SU 811 053 – North of Portsmouth Rd

This was a large mill and, although it had a roller mill system, it was demolished in 1922. This is an ancient site referred to in 1283. *Kelly's* directories give an indication of the occupancy of the mill during the 19th century when Thomas Gatehouse took over in 1851 and later in 1893 he is mentioned in an article in *The Miller* when a 2-sack roller mill system was installed. Turner manufactured the new system with the power provided by both a 10hp steam engine and the overshot waterwheel, 11ft 4in in diameter by 12ft wide. Samuel Gatehouse took over in 1905 and continued until 1918, being followed by W. Bignall, but he had only a short tenancy for the mill was demolished in 1922.
Its position was significant, as a sale advertisement in the Brighton Herald in March 1816 made great notice of the fact that it was situated near to the turnpike road leading from Chichester to Portsmouth. The mill was served by an elongated millpond, bisected by the railway in 1847, and even as late as 1897 the Ordnance Survey map shows an increase in the size of the millpond to the north of the railway line. Today the millpond exists together with some odd pieces of brickwork in the partial remains of the wheelpit. A new housing estate is being built on the site with the millpond retained as a water feature.

CUT MILL *Chidham*
Tributary to Bosham Channel – SU 798 054 – West side of road leading north from village

Cut Mill had been converted into a house by 1924 after milling ceased in 1921. Following the closure, the machinery, including the waterwheel (scrape marks indicate a diameter of 12ft) was removed and a dummy wheel fitted for aesthetic reasons.
Although this is an ancient site, little is known until the publication of *Kelly's* directories. Between 1882 and 1897, the occupier was John Merrett followed by Francis Whybrow, with Amos Wakeford the last recorded miller.
The mill was of three floors of brick, stone and timber powered by a small stream and millpond that issued into the nearby Chichester Harbour.

EAST NEWELL MILL *Hambrook*
Ham Brook – SU 787 079 – By nurseries

East Newell Mill has gone and only the landscaped millpond identifies its position. The last documented reference appears in *Kelly's* 1874 directory with Francis Wright the occupier, but little is known about the mill except that it was demolished in 1913.
There is a reference in 1780 and, from about 1800, it worked in conjunction with Chidham Windmill. According to a fire insurance policy of December 1805, Benjamin Hay was the occupier while, a year later according to the *London Gazette*, the partnership between Hay and Henry Hobbs was dissolved with the former continuing. William Ripking had taken over by 1821 and was listed in the 1840 Tithe Apportionment. By 1845, John Wyatt had taken over and in *Kelly's* directory of that year, he is described as a farmer, miller and coal merchant. The mill remained in the tenancy of the same family until finally closing towards the end of the 19th century. It was a small brick building demolished in 1913 while the millpond was later used for growing watercress, but is now overgrown, and the disappearance of East Newell Mill is complete.

EMSWORTH TOWN MILL *Emsworth*
River Ems – SU 751 058 – North side of Queen Street

This is a large brick built 4-storey mill erected in 1897 and driven by waterpower from the River Ems. The previous mill was destroyed by fire in August 1896 according to reports in *The Miller* in September and November respectively. The fire completely destroyed the site, leaving only the bare walls which were so badly damaged that they had to be pulled down. This mill had been recently fitted with a 2-sack roller plant and the fire broke out while the staff were working overtime; the damage being estimated at £5,000. The first of the two mills in existence here during the 19th century was auctioned for sale in 1820, as the *London Gazette* reported:

'To be sold by auction on the 26 May 1820 at the Black Dog Inn, Emsworth under the commission of bankruptcy issued against Thomas Booker, late of Emsworth - miller and corn merchant. All that capital freehold water corn mill situated at Emsworth with 3-pairs of stones, smutter, with spare running gear. Mill is grinding 15 to 20 loads of wheat per week. A considerable sum of money has been expended in putting the mill and machinery into complete repair.'

Pigot's 1832 directory records that Messrs Clarke & Hellyer were occupiers and were also described as grocers and tea merchants, but this partnership was dissolved in 1853 at a time when they were also maltsters, ship owners and coal merchants. Following the death of Charles Barnham in 1891 the mill was again for sale. The mill, mentioned in 1820, had been enlarged over the years and, according to the sale particulars, it contained 4-pairs of stones and had recently been fitted with the most improved machinery, driven by a 'Little Giant' turbine.

At the time of the 1896 fire, Messrs Chatfield & Whetton who rebuilt the mill owned the site, but they had left by 1900. The sale particulars list 2-pairs of stones along with the modern steel rollers and state that a 25hp oil engine was in use. The mill was working in 1957 under the control of Leigh Thomas & Co (who were also working Old Slipper Mill in conjunction) and were described as provender millers and grain merchants, with the mill entirely powered by electricity, working between 10am and 4pm.

The mill is used as an engineering factory and, apart from the turbine, all milling machinery has been removed. The planning consent for the conversion stipulated that the legend 'Old Flour Mill' should be retained along with the cast iron window frames. At the back, bordering the car park, is the distinctive outline of an overgrown millpond, shown clearly on the 1870 Ordnance Survey 25" map. The mill forms an attractive frontage of Queen Street overlooking the harbour.

FRESH MILL *Fishbourne*
Tributary to Chichester Harbour – SU 837 046 – At the end of Mill Lane

This mill was near to the foreshore of Chichester Harbour, but despite its position, it was never a tide mill, as a spring fed millpond provided the necessary power to a site established here for centuries.

As early as the 15th century, it was known as 'Fresh Myll', when it was one of three mills in the area, belonging to Syon Abbey in Middlesex. In 1555 it was mentioned in the will of John Fenner of Amberley who had become lord of the manor after the Dissolution. William Souter was the miller at the turn of the century, but later, according to the 1840 Tithe Apportionment, Thomas and James Hellyer had taken over. A 1867 directory lists both Farne & Fever, while *Kelly's* 1870 directory lists only Charles Farne. The last mill building was built following a fire in 1917, and was a modern 3-floor mill of brick and concrete (with an internal breastshot waterwheel), but a later photograph shows that it was extended to twice its length on its western side. The rebuilt mill occupied by F Sadler & Co. (also at Westhampnett Mill) comprised of electrically driven machinery and produced pig food in a cube form. This venture lasted until 1928, after which the building lay derelict. The empty building was renovated in 1945, and was used by 'Golden Wheat Products Ltd' until closure in1954, four years after which the building was converted to flats. A long and elongated millpond fed the mill and, as two small streams supplied it, water storage was not a problem though according to records, a steam engine was in use here in 1883. The mill was only some 300 yards from the Old Salt Tide Mill, and could only function when the sluice gates of the tide mill were left open to allow waste water to escape. The owner of the tide mill, James Shepherd, was rather an unhelpful fellow and at times refused to open the gates. To overcome this problem, according to an entry in The *West Sussex Gazette* and County Times of 29 October 1857, a windmill was bodily moved from Rustington to Fishbourne for Henry Fever to use. In fact, the post mill continued in use until June 1898, when it was demolished. A more serious threat to the watermill's existence occurred in 1874, when waterworks were established nearby which drew water from the two streams that fed the millpond.

Now, the external appearance of the 'mill' has been altered out of all recognition with the water diverted, and it is difficult to visualise sailing ships from Southampton delivering grain to this mill in the last century.

LUMLEY MILL *Westbourne*
River Ems – SU 752 064 – West side of lane linking Emsworth with Westbourne

Lumley Mill was situated on the Sussex side of the River Ems, but now comprises of just some brick footings, odd pieces of machinery and a dried up millpond.

The mill was erected in 1760 when, as part of the Stansted estate, it was in the ownership of James Lumley MP, following which Richard Barwell purchased the estate in 1778. He sold it in 1802 to Edward Tollervey who was a businessman with many commercial interests in the Portsmouth area. He was a man of some financial substance for he built the pseudo gothic mill house together with an extensive range of outbuildings that included large bread ovens and pigsties. Flour was produced in great quantities at a time when both bread and biscuits and fresh meat were in great demand by the thousands of troops stationed at Portsmouth, and by the naval contracts. By 1815, the Napoleonic wars were over and the Admiralty decided to construct their own mills and bakeries in Gosport.

This was the end for Tollervey and owing a great deal of money to his creditors, he was declared bankrupt in 1820. Tollervey was so distressed by the loss of his business that he was seen as a road sweeper in Fleet Street in London some years after.

The sale advertisement for Lumley Mill, that appeared in the *Sussex Weekly Advertiser* of 26 August 1821, gives an insight into the size and scale of the business here. It was described as a mill close to the quay at Emsworth, on a stream of uncommon and unceasing power, having stores that could accommodate 500 loads of grain or flour, an extensive malthouse, threshing machine, biscuit manufactory, three large bread ovens and a large residence.

According to the 1842 Tithe Apportionment, Edward Harker was owner with William Shean the miller and following Shean's death in 1860, James Terry took over and the mill remained with the family until 1915, his nephew J. Alfred Terry being the last miller. The mill and most of the outbuildings were destroyed by fire on the 24 May 1915, but fortunately the mill house was saved.

Thereafter the site was never re-used and what was left of the mill and the outbuildings lay disused and ruinous for many years. There is now little left on the site, but the outstanding mill house remains which has the legend 'Lumley Mill' emblazoned on its front wall. The outline of the millpond is just discernible and in the wheelpit are the remains of the iron overshot waterwheel, 10ft in diameter by 7ft 9in wide and part of the 8ft diameter iron pit wheel. On the northside is a detached small pinion wheel that once must have driven ancillary machinery. The mill was powered by the River Ems which was widened to form a canal type basin on the north side. The site is found on the west side of a lane that connects Emsworth with Westbourne and a little further on are two old cottages that had connections with the mill. However, the traffic noise generated from the close proximity of the A27 by-pass has shattered the tranquillity of this site.

MID LAVANT MILL *Mid Lavant*
River Lavant – SU 856 086 – North of Sheepwash Lane east of village

The mill is situated at East Lavant and, during the Domesday Survey, was worth 7s. Little is known about the mill other than, according to the 1851 Tithe Apportionment, George Parsons occupied the mill, and that it was referred to in an 1867 directory. The mill had worked in conjunction with East Levant windmill until the latter was pulled down in 1843.
The brick built 3-storey mill has been converted to a house with little indication of its past.

NEW FISHBOURNE SALT MILL *Fishbourne*
Chichester Channel – SU 837 044 – South of Mill Lane

This tide mill was almost adjacent to the Old Salt Tide Mill and obviously worked in conjunction with it. According to the Ordnance Survey 25" map published in the 1880's, the outline of the pond is clearly shown with the site marked 'Old Mill Pond (Tidal)'
Denis Sanders referred to this mill as 'Barton's Mill' and in a visit in July 1957, he stated that some masonry survived although not in a recognisable form.
Nothing is known about the mill and according to Mr Hawksley it was demolished in 1763.

NEW SLIPPER MILL *Emsworth*
Tidal mouth of River Ems – SU 754 053 – South of Slipper Mill

Little remains of this tide mill erected by Mr Hatch in 1867 following a dispute with Mr Byerley, owner of nearby Old Slipper Mill, who argued that the tail water interfered with the running of his mill.
The mill only survived for just over thirty years until it was destroyed by fire in 1879 and it is now difficult to establish its position. In 1959, it was reported that the millpond was filled with large tree trunks, left by a timber company to season in salt water. Although there are still some traces of old brickwork to be seen, part of the former millpond has been filled in while the rest is used as a yacht basin and yard.

NORTHBROOK MILL *Funtington*
North Brook – SU 808 077 – East side of West Ashling to Funtington Road

This was a large rectangular brick built building of a size comparable with other mills that stood on the same river. The last mill was erected, according to a stone tablet 'D.D.R,' in 1721 and refers to Robert and Ruth Drinkwater, while later in 1739, the *London Gazette* records a commission of bankruptcy against Woodruff Drinkwater who was also in control at Sidlesham Tide Mill.

According to the 1838 Tithe Apportionment, the occupier was John Smith, while *Kelly's* 1870 directory refers to Smith & Wood, with the latter continuing as owner/occupier until he left the mill to Henry Goodyer in 1874, in whose family it remained thereafter. Milling ceased in 1928 following an unfortunate accident with the pit machinery. The pit wheel was being recogged when the millwright was called away. During his absence torrential rain filled the wheel buckets and set it in motion, and the momentum of the waterwheel broke the secured pit wheel in two and Mr Goodyer did not consider the repair expenses to be justified. A dynamo installed in 1915, continued working for some time afterwards.

An inspection by Sydney Simmons in September 1946 gives an insight into the machinery fitted to the mill. The waterwheel was 12ft in diameter by 7ft wide made of iron and overshot and was inscribed 'W. Thomsett, Godalming 1882' (Simmons surmises that the previous wheel was wooden and undershot and the tall iron columns supporting the new overshot wheel and pentrough would suggest this). On the pit floor the broken iron pit wheel lay discarded and on the iron upright shaft was the 3ft 6in diameter iron wallower and 10ft 6in diameter iron spur wheel, attached to which were four stone nuts mounted on iron bridge trees with Jack Ring lifts. There were 4-pairs of stones in round tuns and a stone lifting crane. On the floor above was a iron crown wheel 6ft 4in in diameter which drove an assortment of line shafting and flour machines, with the sack hoist operated by two pulleys working in an inclined wooden frame.

When Denis Sanders visited the mill in July 1957, the machinery had been removed apart from the waterwheel, which was in good condition. Pigs were housed on the ground floor with chickens on the floor above. Northbrook Mill was demolished soon after to make way for a modern building built at right angles to the mill house with the waterwheel and pentrough removed for scrap at the same time.

This is an idyllic site with a millpond at the back (a third of which has been filled in) used for breeding trout. Although there are a few foundations identifying part of the mill, the disappearance of Northbrook Mill is complete.

NUTBOURNE TIDE MILL *Nutbourne*
Ham Brook – SU 776 050 – On foreshore south of village

This tide mill stood on the edge of Thorney Channel, south of the village of Nutbourne. It was a large mill described in a sale notice, in *The Miller* of 5 May 1884, as a 'powerful brick and stone built freehold Corn Tidal Mill, driving six pairs of stones, known as Nutbourne Mill, with excellent machinery.
Access was by road or sea, and an earlier sale notice in the *Sussex Weekly Advertiser* of February 1791, stated that there was sufficient water for vessels to unload at the mill door. There was also a millpond of over 20 acres.

It is not known when the site was established, but a tide mill was operating in June 1723, according to a sale notice in the *London Gazette*, when it was capable of grinding 400 loads of wheat per year and there was a very good kiln on the site.

From at least 1845, the mill was in the occupation of the Wyatt family, with Francis taking over from John in about 1874, until he was superseded by Fred Wyatt until 1884 (also at East Newell Mill). The mill continued to at least 1890 with James Hackett the last miller, after which all references cease. There is no sign of the mill site and the millpond has become pasture land, although the embankment is just discernible, while the Ham Brook issues into the sea at a point where the mill probably stood.

OLD SALT MILL *Old Fishbourne*
Fishbourne Estuary – SU 837 044 – South of Mill Lane

The site of Old Salt Mill can be found 300 yards south of Fresh Mill in an area of endless mud flats at low tide. It is not known when this tide mill was established (it being shown on Budgen's 1724 map) but, by all accounts, it was a very small building.

During most of the 19th century, the mill was in the occupation of the Shepherd family with James being the last occupier according to *Kelly's* 1911 directory. The exposed position of the mill, along with the ever present threat of flooding, was a normal problem associated with tide mills.

According to a report on the mill by Rex Wailes in November 1940, only the foundations and brick culverts remained. It was built of a mixture of red coloured stone of various sizes under a tiled roof, worked by a single waterwheel. During flooding, the embankment (which then and now provides a path to the site) was sometimes under water.

There is little to be seen today but its position is identified by a scattering of stonework at the end of the causeway.

OLD SLIPPER MILL *Emsworth*
River Ems – SU 745 055 – At end of Slipper Mill Lane

Old Slipper Mill was a tide mill that stopped operating in 1949 according to Denis Sanders. It powered 2-pairs of stones from a single wooden undershot clasp arm waterwheel, 9ft in diameter by 8ft 6in wide. Rex Wailes is of the opinion that the mill dates from 1735, but it was certainly the last working tide mill in the county. It was a three-storey brick built mill with a corrugated iron roof. The wooden waterwheel drove iron wheels and a wooden upright shaft and 2-pairs of stones (originally 3). From the crown wheel there were drives to a combined kibbler and crusher, grindstone, shaker screens with blower, and an assortment of ancillary machinery with the sack hoist of special interest.

According to *Kelly's* 1845 directory, Thomas Byerley who was also working Westbourne Mill in conjunction, occupied the mill. Fred Byerley had taken over by 1895 and it continued in the ownership of the same family until at least 1934. In 1911, James Thomas & Co leased the mill, they in turn being replaced by in 1922 by Thomas Gater and Bradfield & Co. (an Isle of Wight based company) who worked it in conjunction with Emsworth Town Mill until at least 1949. An article in the *Countryman's Diary* of 1941, stated that Frank Burgess had worked at the mill for 56 years, ending up as the manager.

In July 1957, it was reported that the frame of the clasp arm waterwheel was still in place, but the machinery had been removed and the adjacent large brick building used for storage. A later report in July 1963, revealed that the mill had gone and only some of the lower walls remained and that the pit floor of the mill had been filled in to bring it level with the adjacent roadway. The mill was powered from a pond fed from by a combination of water from the River Ems and the ebb tide flowing out to Emsworth Channel.

The base walls of the mill survive and the adjacent former brick built granary has been converted into residential properties overlooking a harbour utilised now for recreational purposes.

OVING MILL *Oving*
Tributary To Aldingbourne Rife – SU 905 049 – Southeast of village

The mill site lies in flat countryside to the south of Tangmere Airfield and was run in conjunction with a windmill situated close by. A sale notice in the *Sussex Weekly Advertiser* of 20 March 1797 refers to a 'Newly erected Water Corn Mill' and a 'newly erected windmill', and again in 1815, the *Hampshire Telegraph* refers to the sale of both mills.

The 1841 Tithe Apportionment records George Broadbridge as occupier of both mills, with Charles Saunders taking control from 1845 until at least 1858. William Stoveld followed from 1862 until 1870, after which all references to both mills cease, and it is thought that they were both demolished soon after.

There is no evidence to be seen to indicate the watermill site but a trackway, through the adjacent dairy farm, leads to a crossing point of a stream where the mill probably stood. The mill house stands as a mute reminder to both the wind and the watermill.

RATHAM MILL *Bosham*
North Brook – SU 812 063 – Adjacent to Ratham Lane

The first reference to a mill is found on Budgen's 1724, and it is not until the 19th century that further information exists. In November 1800, the *Sussex Weekly Advertiser* advertised the mill for sale, occupied by Mr Creswell, while in March 1807, the same newspaper reported on a fire here:

'On Saturday morning, about 1 o'clock, a fire broke out in a kiln house adjoining the Water Corn Mill at Ratham, near Chichester, which raged with such fury that mill, with its contents upwards of 50 loads of wheat and flour, totally consumed before assistance could be procured. The mill was the property of Lady Louisa Lennox, mother of the Duke of Richmond, and was used by William Smith, whose loss is estimated at upwards of £2,000, neither the mill nor the stock being insured'.

William Smith was the proprietor of the rebuilt mill but, in March 1818, he was made bankrupt according to a report in the *Hampshire Telegraph*. The newspaper relates that the mill had a millwright's shop, wagon shed for six horses, a carter's lodge, a residence and that the mill ground 14 loads of wheat in the driest weather. William Smith Jnr carried on as miller until the mill was sold. William Jeffery was the occupier according to the 1839 Tithe Apportionment, and he continued to at least 1890 (also working West Ashling Mill before 1870). John Heaver took over in about 1909, according to *Kelly's* directory, until replaced by his son John in 1929. In 1935 it was working as a grist mill using the newly refurbished waterwheel.

When Peter Davies visited the mill in 1946, he reported that the waterwheel was only used for hoisting and for driving an oat crusher and that the mill contained a good deal of electrically driven machinery. In July 1957, it was reported that the 3-pairs of stones had been removed although the waterwheel was still used for hoisting. Later in 1978, a large modern electrically powered grain mill was built almost adjacent to the old mill, which reverted back to a store. The iron overshot waterwheel, 9ft in diameter by 8ft wide, and the pentrough, were manufactured by 'R. Chorley Midhurst'. The turbine (used for generating electricity) remains in-situ while inside the mill, the upright shaft remains, boarded on one side, through which an iron bar protrudes, which connects with the gate of the pentrough. There is also a 9ft diameter pit wheel, 8ft 6in spur wheel and a wallower, all in reasonable condition, together with the remains of a worm elevator. On the stone floor is an electric generator that was driven by a belt from the turbine house. There is a Eureka No.2 separator and a bolting machine manufactured by Turners, while on the floor above are the remains of a sack hoist. It is a large brick building of 4 floors and apart from the poor condition of the roof purlins, it is structurally in a good condition. The modern power mill has recently been demolished and the future of the old mill is uncertain. Ratham Mill was once one of three large watermills working in the area, but its survival was due to the owner's perseverance in keeping up with modern milling methods, until it succumbed to the inevitable.

RUNCTON MILL *Runcton*
Pagham Rife – SU 880 022 – In Mill Lane

The is without doubt an ancient site, mentioned in the Domesday Survey, and one of the first references appeared in the *Sussex Weekly Advertiser* of April 1795, when the mill is described as newly built with 2-pairs of stones. In the sale particulars, reference was made to a nearby windmill, also newly erected. Another sale notice in 1808 stated that the mill was capable of grinding 8 loads with french stones and that it was working in conjunction with the nearby smock windmill, with Thomas Brewer the proprietor, (who had dissolved a partnership with William Boniface according to the *London Gazette* of 8 April 1808.) James Hunt was the occupier according to the 1850 Tithe Apportionment, he having taken over from John Shepherd, and remained until to at least 1873, when he gave up. There was a succession of millers afterwards, according to *Kelly's* directories i.e. 1878 - Earl Wedge, 1882 to 1903 - Hodson Bros., 1905 - Fred Hodson, 1911 - West Sussex Milling Co. (steam only), 1913 - John Baxter. It appears that the mill stopped working in 1915 and was converted into a house before the last war, all machinery, including the breastshot waterwheel, having been removed. The mill forms part of an attractive group of buildings brick built to four floors adjacent to an extremely old mill house, while a large millpond exists. The water still pours through the wheelpit, which is covered by an extension to the kitchen, outside, the engine house chimney survives.

SIDLESHAM TIDE MILL *Sidlesham*
Pagham Harbour – SU 862 973 – South side of Mill Lane

This was a large tide mill situated at Pagham Harbour, erected in 1755 by Woodruff Drinkwater, that replaced an earlier timber framed structure shown of the survey on the Sussex Coast in 1587.

The mill, erected by Drinkwater, was a large three bay 5-storey structure of Caen stone. It had three waterwheels driving 8-pairs of stones and could grind a load of wheat in one hour, and a report in the *London Gazette* of 1772, reveals that tidal power was derived from a 50 acre millpond. It was also reported that sailing vessels could unload and load at the mill door.

As early as 1664, there were proposals to reclaim the harbour to agricultural land but it was not until 1852 that a detailed survey was carried out by the Harbour Department of the Admiralty. This revealed that the harbour was nearly two miles broad and long and recommended enclosure and reclaiming the land, which caused great concern to the mill owners and tenants. At this time the harbour was still in use for trading and in that year sixty-eight boats, carrying on average 22 tons each, delivered coal and grain to the mill with flour collected for distribution. Perhaps the mill owners had perceived the financial threat to their existence, as a beam engine had been installed here a few years previously.

From a photograph taken at the turn of the century, a large boiler chimney dominated the mill, but despite the steam assistance, the mill closed down in 1900 and was demolished in August 1919, with the materials used for local housing.

According to a sale advertisement in 1772, Drinkwater was still the owner with a Mr Cooke the mill manager. In 1777 a fire insurance policy indicated that James Parsons, a Portsmouth merchant, was the owner of the mill, while Michael Kingsford took over from 1782 until 1804. The 1848 Tithe Apportionment records that William Sharp was the occupier, he being succeeded by Kiln & Clarke who stayed until 1862, after which James Clayton took over.

Only some brick footings remain to indicate the position of this large tide mill and even the embankments of the millpond have gone, with Pagham Harbour now a nature reserve. The mill house and cottage remain in this rather isolated corner of West Sussex.

WEST ASHLING MILL *West Ashling*
North Brook - SU 807 074 – Junction of West Ashling Road and Watery Lane

This appears to be an ancient site that was converted into paper making in 1823. After being used for flour milling for centuries, a new building was erected for paper production. Although it was producing coarse paper in great quantities, the mill changed hands on a regular basis until it stopped in 1851. Soon afterwards the building was converted back to flour milling. Contemporary photographs show that a windmill was erected on the top of the mill and Martin Brunnarius, writing in his book *The Windmills of Sussex*, states that it was erected in 1861. Unlike similar developments at Hooker's Mill at Twineham and Bishopstone Tide Mills, this windmill was used for driving millstones (3 in total) and was independent of the power source of the watermill. The windmill was mainly a wooden structure, open and supported on quarter bars on the roof of the mill. At the apex of the trestle was a circular ring, enclosing the iron upright shaft (within an iron casing), and had four patent sweeps and worked into the early years of this century (by Hackett & Son). The watermill also had 3-pairs of stones in a layshaft arrangement, originally powered by a wooden clasp arm breastshot waterwheel 12ft in diameter by 10ft wide, subsequently replaced by an Armfield turbine in 1930. The 3-pairs of stones are on a stage and were driven by belts and pulley wheels (one above the other) from the turbine shaft. There are no such traces of the windmill, the remnants of which were pulled down in 1955. *Kelly's* directories record that Walter Jeffery was using the mill until at least 1890.

The windmill stopped working when George Daniel's was the miller, but from 1899 Hackett & Sons took over and remained until closure in 1941, with Francis Hackett the last miller when it was only producing cattle feed.

After the closure, the mill fell quickly into disrepair and in 1946 part of the roof collapsed with most of the woodwork rotten and it remained as such until 1978. However, the complete restoration of the old mill building has taken place and it has become a house and office. For once, the conversion has been tastefully carried out with the turbine driven machinery, sack hoist and flour bolter retained. The mill is brick to three floors under a slate roof with parts of the ground wall being 3ft thick. The old miller's house at the back of the mill has been demolished.

West Ashling Mill was one of five watermills on this small stretch of river and is hidden away, with its idyllic millpond, in a quiet and attractive corner of West Sussex.

WESTBOURNE MILL *Westbourne*
River Ems – SU 807 074 – West side of River Street

This is a substantial brick building and the highest mill on the River Ems. It is undoubtedly a Domesday site as Westbourne was a trading centre from early times, with the mill working until 24 October 1933.

Although reference was made to the mill in the 1803 Defence Schedules, it is not until the publication of the *Kelly's* directories that some of the millers are known. Thomas Byerley (also at Old Slipper Mill) was here from 1845 until at least 1862 and after a brief occupation by James Guyatt Jnr, he relinquished it to Henry Stewart and then James Challen. According to a report in *The Miller* of 3 March 1890, the mill contained 5-pairs of stones (3 water, 2 steam).

The mill building comprises of three floors and a basement built in red and blue brickwork under a tiled roof. It had a 12ft diameter iron waterwheel with wooden rims, iron wheels and a wooden upright shaft. In its latter working days the mill contained 3-pairs of stones, a grindstone, dynamo, oat crusher, two silk flour machines, smutter, elevators and a triple pump for water for the village. The mill, fronting River Street, was converted into a house in 1958 and is devoid of any machinery with access strictly private.

WESTHAMPNETT MILL *Chichester*
River Lavant – SU 877 060 – West side of Madgwick Lane

The large five storey brick building was erected in 1906 to replace an earlier mill burnt down in October 1904. It was totally engine driven and used a $1^1/_2$ sack Turner roller milling system which, during WW1, was upgraded to a $2^1/_2$ sack system. Prior to the fire, a beam engine was used when the water level was low, purchased in Glasgow and delivered to Dell Quay by boat. Apparently the beam engine was not destroyed by the fire but was out of use by then. A gas engine installed in 1907 was replaced by a Vickers Peters 75hp semi-diesel engine but it is not known when the mill stopped working, but *Kelly's* 1934 directory is the last reference to the mill.

The mill was in the ownership of William Knott according to a fire insurance policy, but in April 1827 Pannell Knott was made bankrupt. There was no recorded miller here in 1845 but by 1851, Robert Sadler had taken over and it remained with the family until 1934, when the mill was being run under the name of A. Sadler & Co.

Until quite recently the buildings were devoid of machinery, with water still flowing through the old wheelpit under the mill where once stood a breastshot waterwheel, but in June 2000, the mill was converted into four residences. The River Lavant that powered the mill, now flows quietly through the site and it is difficult to envisage the capabilities of this small river that caused the extensive and devastating flooding of Chichester in 1985.

Birdham Tide Mill in 1937 just before closure (DW)

Aldingbourne Mill with its boiler hose (VU)

Waterwheel at Aldingbourne Mill in 1956

Sidlesham Tide Mill in the early 1890's (VU)

The extensive Sidlesham Tide Mill seen across Pagham Harbour (AS)

Overshot waterwheel at Ratham Mill (DW)

Ratham Mill in 1994 showing signs of disrepair

Waterwheel and turbine house at Ratham Mill in 1994

Bosham Mill in 1891 with its disused waterwheel (BLSL)

The millrace to Bosham Mill in 1910 (VU)

The external waterwheel at Birdham Mill just after closure in 1936 (AS)

Fresh Mill at Fishbourne in the early 1900's (VU)

Fresh Mill in 1951

The unusual pseudo-gothic Lumley Mill House in 1993

Northbrook Mill and pond embankment in 1936 (DW)

The waterwheel at Northbrook Mill in 1936 (DW)

Northbrook Mill and house in 1939 (VU)

Ruins of the Old Salt Mill in 1939 (AS)

Old Slipper Mill across the millpond in 1959

The stone floor at Old Slipper Mill (AS)

Emsworth Town Mill as an engineering factory in 1993

An idyllic scene at Runcton Mill in 1994

West Ashling Mill at work in 1910 (DW)

Removal of the watermill at West Ashling Mill (DW)

The remains of the windmill at West Ashling Mill in 1954 (AS)

Layshaft drive machinery at West Ashling Mill (AS)

The disused West Ashling Mill with the turbine extension in 1978

The solidly constructed Westbourne Mill in 1936 (DW)

The abandoned wheelpit at Westbourne Mill in 1957 (DW)

Westhampnett Mill as a factory in 1951

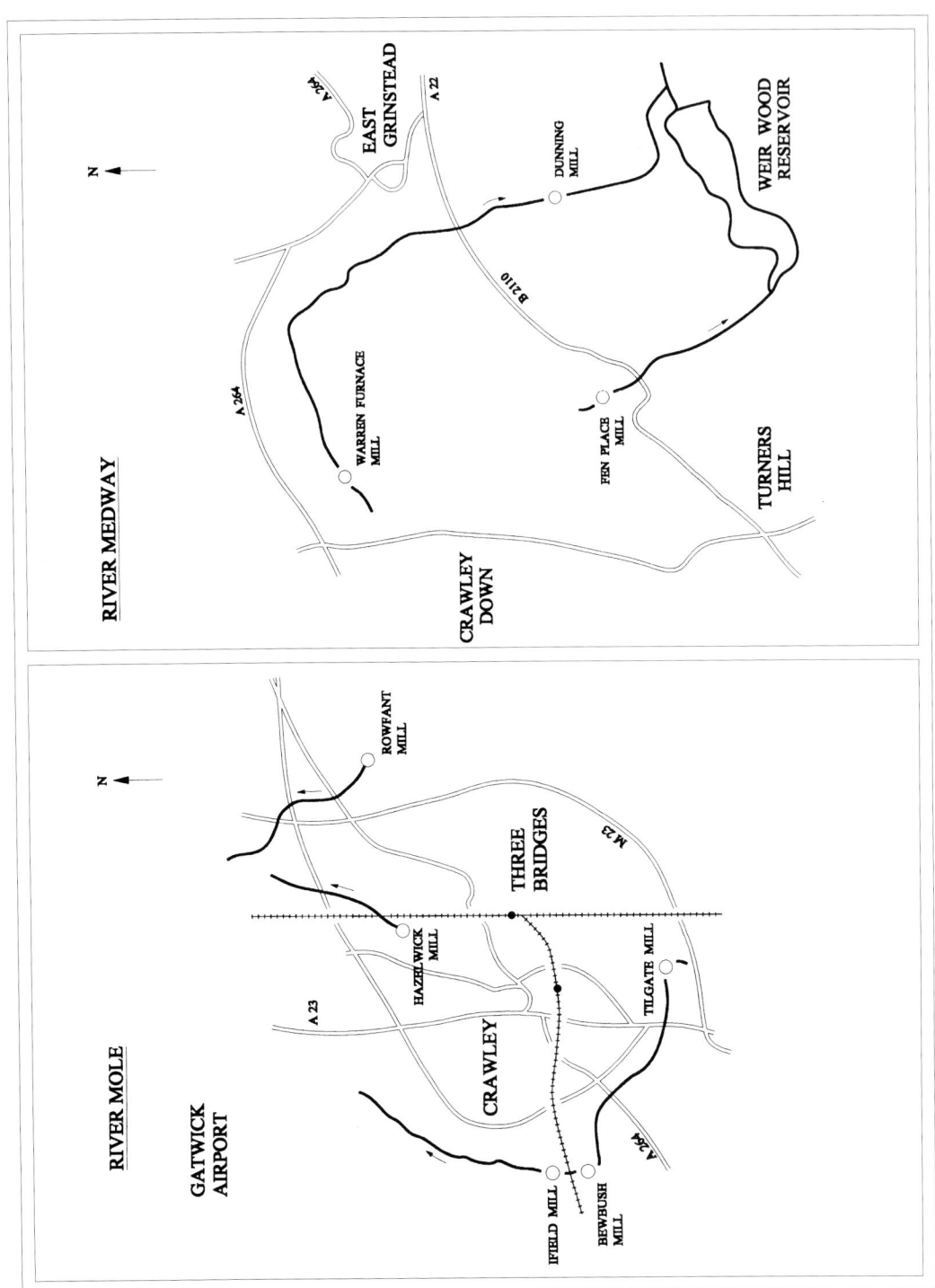

RIVER MOLE & RIVER MEDWAY

BEWBUSH MILL *Bewbush*
Tributary to River Mole – TQ 245 362 - ½ mile south of Ifield Mill

Bewbush Mill was once marooned in open countryside to the south of Ifield. The advance of Crawley New Town has drastically altered the surrounding area, although the mill site has remained untouched so far. There was once an iron furnace working here in 1569, utilising a shallow but large millpond, but this was out of use by 1642, reportedly due to lack of wood.

As with several other sites in Sussex, the extant millpond was subsequently re-used to power a corn mill, which had been established by 1787. This was only a small 2-pair mill according to a lease advertisement in the *Sussex Weekly Advertiser* of March 1830.

The mill was marked 'Disused' on the 1874-5 Ordnance Survey 6" map, and being a small country mill, it could in no way compete economically with the modern trends of corn milling. The mill stood derelict for some time and according to Denis Sanders, was finally demolished in June 1931. A photograph taken at the turn of the century depicts a small brick building forming part of the pond embankment, with entry directly onto the stone floor, with a by pass arrangement on the southern side. During the demolition of Mill Farm Cottage in 1976, a millstone was uncovered which was removed for conservation and use at Ifield Mill.

The search for the site of Bewbush Mill is certainly not easy but the pond embankment remains, with some brickwork identifying the site.

DUNNINGS MILL *East Grinstead*
Tributary to River Medway – TQ 393 368 – Adjacent to West Hoathly Rd

This was a small watermill situated on the southern outskirts of East Grinstead and, apart from a reference to a John Cole here in 1597-8, little is known about its early history.

The 1803 Defence Schedules lists Peter Everest as miller and refers to the mill as 'Framepost Mill'. James Turner had taken over by 1839 according to *Pigot's* directory and remained until at least 1878, and it was during his tenancy that the mill became known by its present name. The last directory entry appeared in 1890 with Luke Streatfield and Luke Godley in partnership and it must be assumed that commercial milling stopped soon after.

In 1938, Ernest Straker, writing in the *Sussex County Magazine*, reported that the millpond had been drained but the machinery left intact (this was removed in 1950) and also that the mill had been combined with the mill house. Both buildings have become a restaurant but there have been quite substantial building alterations. A section of the wooden upright shaft has been retained and resited to an adjacent room, but the small tributary to the River Medway still passes through two brick lined channels under the mill, but of the millpond, there is no trace.

FEN PLACE MILL *East Grinstead*
River Medway – TQ 363 366 - 1½ miles southwest of town centre

This is the highest mill site on the River Medway, set in a picturesque location with the mill, millpond and mill house forming a most pleasant scene.

This does not appear to be an ancient site, although Ernest Straker refers to 'Bishes Mill' here in 1598, according to an article in the *Sussex County Magazine* of 1939. The 1803 Defence Schedules refer to Richard Heather as the miller, while later John Stanbridge (also at Warren Furnace Mill) was the occupier between 1832 and 1870. The 1873 Ordnance Survey 25" map marks the mill as 'Corn' indicating its use as a provender mill, but afterwards there were several millers in occupation i.e. 1874 - John Miles, 1878 - William Stevenson, 1882-87 - David Dadswell, 1890-1900 - Hugh Markwick, after which references cease. The mill, like many others, continued working on a casual basis but after WW1 the machinery was removed and the building became a farm store. Little is known about the type of machinery apart from the fact that in 1903, the overshot waterwheel was wooden, of the clasp arm variety, and in good condition.

The building was constructed in two stages, with that part nearest the millpond of much greater age that the extension on its southern side. As the highest mill on the River Medway, water problems must have been a constant worry to the millers, but the large millpond must have eased the situation. The mill has been converted into a house and only its name 'Fen Place Mill' along with the millpond, indicates the former use of this site.

HAZELWICK MILL *Crawley*
River Mole – TQ 287 378 - In Hazelwick Mill Lane

This mill stood in open countryside north east of Crawley town centre, until the area succumbed to the building of Crawley New Town after WW2 but, by then, the mill had closed and was disused.

There is a reference to 'Hassellwick Mill' with pond, floodgates and mill dam, in the sale particulars of the manor of Worth in 1643, but it is not until the beginning of the 19th century that further details are known, for in the 1803 Defence Schedules, the miller, William Tidy could supply one sack of flour daily.

Throughout the rest of the century the mill was in the control of the Caffyn family until closure. During their occupation they also used Copthorne windmill in conjunction for forty years. Towards the end of the century the competition from the large steam powered mills led to the end of flour milling at Hazelwick Mill. Unperturbed, the Caffyn's set about remedying the situation by building their own steam mill in Three Bridges, according to an article that appeared in *The Miller* of 11 January 1897;'Messrs P Caffyn & Sons. Three Bridges Mill, Sussex. These fine new buildings contain a complete Roller Mill Plant on the Simon system. The whole established was erected in the latter part of 1895 and early in 1896, and is the property of Messrs P Caffyn & Sons, an old established firm of millers whose business was originally carried on at the Hazlewick Mill and at Horley Mill where the old mills are still working. The new mills were set to work in June 1896.'

It appears that Hazelwick Mill never worked after the Caffyns left the mill and it become disused and in 1939 it was described as a 'dilapidated ruin'.

The mill was a small brick and weatherboarded mill, with a tiled roof, and formed part of an artificial pond embankment, but what is surprising, is that it had two overshot waterwheels, approximately 12ft in diameter, working in tandem. An inspection of the mill in 1939 revealed that only the iron shaft of the front wheel remained, which had an iron box with the date '1861' inscribed upon it. The rear wheel was complete but in a poor condition. Each waterwheel was fed by its own pipe, and as the overflow also had a pipe outlet, so the wheelpit must have been rather congested.

Hazelwick Mill is the second highest watermill on the River Mole and subsequently had a large millpond. The mill was adjacent to the London-Brighton railway about 2 miles north of Three Bridges railway station, where the pond extended to, and where the floodgates controlling the water supply from the River Mole were situated.

There is nothing to be seen, as the millpond has been replaced by a supermarket car park, and only the road name Hazelwick Mill Lane perpetuates its memory.

IFIELD MILL *Ifield*
Tributary to River Mole – TQ 245 364 - Adjacent to a track leading south of Rusper Road

In 1973, Crawley Borough Council started to acquire land in Broadfield Vale for housing development, part of which included Ifield Mill. At the time the mill was in a derelict and ruinous condition, devoid of machinery and situated adjacent to a 12ft high pond embankment that was also in a poor condition. Since 1974, in a countryside of many empty mill sites, Ifield Mill has been painstakingly restored and in the summer of 2000 will once again be milling flour, about 80 years after the mill stopped working.

This is without doubt an ancient mill site, first mentioned in 1274 when it belonged to Rusper Nunnery, and it once stood in a densely wooded forest that was later decimated when the local forges and furnaces were in operation, as vast amounts of charcoal were needed. The blast furnaces at both Bewbush and Ifield were working in 1574, with Roger Gratwick the tenant. The decline took place in the C17th with both forges reported as derelict in 1653.

It is not clear when the corn mill was erected, but it was common practice to take advantage of the large expanse of water that invariably survived at the cessation of the iron industry. Thomas Middleton was the first owner of the mill and a stone tablet incorporated into the present mill building bears the inscription 'M.T.M' and the date 1683, but nothing of this mill survived the subsequent alterations to the site. It appears that the new miller was William Garton who, being a Quaker, suffered religious persecution and was later imprisoned in Horsham jail. However, his family continued in occupation of the mill until about 1742, when Leonard Gale purchased the ownership of the mill. He sold it to John Leake in 1759 and ran the mill with various business partners. In 1809, Abraham Goldsmid purchased the mill, but he died in debt a year later and for the next seven years the mill was in the hands of his executors. The exact date of the rebuilding of Ifield Mill is not known but both Straker and Simmons give 1817, but do not quote their source. The new mill was purchased by Thomas Durrant, the first owner-miller, for £1,800. Durrant had moved to Ifield from Merstham Mill where he received compensation from the owners of the nearby stone mines, when they cut off the supply of water to the mill. Difficulties with maintaining a sufficient head of water at Ifield started to become apparent, which was not helped when the 17 acre millpond was bisected by the Crawley – Horsham railway, first on an open trestle bridge and later by an embankment. Also, in 1837, James Bristow constructed a windmill on Ifield Green in direct competition with the watermill. Richard Harding was working as a tenant-miller in 1875, before purchasing the mill 10 years later. The mill continued using steam engine assistance when necessary, up to the First World War when labour shortages caused the mill to close down commercially.

The mill house in 1934 was described in a sale catalogue as a 'gentleman's residence with a picturesque disused watermill' and it remained as such until the site was purchased by the local council.

In 1974 a formal approach was made by the Crawley and Mid Sussex Archaeological Group to restore the mill subject to a feasibility study being carried out. The Council granted permission to the group (which became known as the Ifield Mill Project). One of the main problems, apart from some structural damage, was that the mill contained little machinery, although the frame and axle of the waterwheel existed. Therefore, a search was mounted to acquire machinery from other mills under the threat of demolition or conversion, where suitable machinery could be salvaged. One such mill was Hammonds Mill near Clayton, which was being demolished in 1975 and with the help of the contractors, the project team were able to remove the cast iron pitwheel, vertical iron upright shaft, wooden wallower and spur wheel, cast iron hurstings, and a cast iron penstock (manufactured by Coopers of Henfield). Smaller artefacts were rescued from Balcombe Mill (then under threat of demolition but now converted into a house). These included many timber fittings and also a set of beam scales that were removed for conservation and later use.

The external overshot iron and timber waterwheel at Ifield is exceptional in that it measures 11ft 6in in diameter by 11ft and is a fine sight, when turning. The millpond was drained in 1977 by the council and this, together with the mill, forms a most picturesque sight, far removed from the description in 1934! The restoration of Ifield Mill is almost complete and represents 25 years of hard work by a loyal band of volunteers. Without doubt, but for their efforts, the mill would not be standing if the vision of its repair had not been realised back in 1974.

ROWFANT MILL *Worth*
Tributary to River Mole – TQ 316 378 – Adjacent to track leading off Old Hollow

Rowfant Mill forms part of an attractive group of buildings that have been converted into houses. During the early 17th century, a forge was operating here, but had reputedly closed by 1664. As with many Wealden Iron sites found in East Sussex, a corn mill was later established to take advantage of the large water supply; in this case, Calder Pond. Ernest Straker is of the opinion that a corn mill preceded the iron forge, but this cannot be verified. The date of the resumption of flour milling is unknown, but the age of the mill would appear to be middle 18th century.

The 1803 Defence Schedules refer to Thomas Stone as miller and later Isaac Sayer was working here, but in the 1850 Tithe Apportionment, Thomas Heasman had taken over and according to *Kelly's* directories it remained with the family until at least 1924. In 1866 William Heasman moved to Coltsford Mill, at Hurst Green, Surrey and added a modern roller mill to Oxted Mill in 1888, indicating that the Heasman's were a prosperous milling family. In 1869, a new iron overshot waterwheel and pentrough, manufactured by J. Morley at Crawley, was fitted with the axle shaft extended to connect with a belt drive.

It is not known when Rowfant Mill stopped working, but it ended its days as a provender mill. Certainly by the late 1920's, the mill was disused, with the machinery removed in 1938, and converted into a house in 1946.

The original brick and weatherboarded mill and the adjacent store remain structurally unaltered, apart from the necessary extra window space on the western flank wall. The wheelpit still survives, but the waste water from the pond now pours unchecked through a landscaped garden. The mill itself is brick to the first floor with three floors of the traditional weatherboarding above, under a tiled mansard roof. As the mill is set into the embankment of the millpond, access into the stone floor could be made from the track passing the front of the mill.

This is a most peaceful site, the mill being a bed and breakfast accommodation, with millstones scattered around the site, acting as a mute reminder of its past use.

TILGATE MILL *Crawley*
Tributary to River Mole – TQ 280 346 - In Tilgate Park

There was an iron furnace north of this site, in operation in 1653, but the first reference to a corn mill was in a sale notice in the *Sussex Weekly Advertiser* of 15 December 1785. In 1793 it had 2-pair of stones, while later the 1803 Defence Schedules recorded that the miller, William Seaton, could provide 3 sacks of flour daily. The last reference appeared in the *London Gazette* on 22 June 1827 when the estate, including the mill, was advertised for sale.

The large millpond is now the only legacy that remains, but a careful inspection reveals the remains of a dried up tail race and some dressed stone blocks which could have been associated with the mill.

WARREN FURNACE MILL *Felbridge*
Tributary to River Medway – TQ 347 392 – Southwest of Furnace Farm Road

This was the site of an iron furnace from 1609 until at least 1653 and replaced by a corn mill soon after. It is marked on Budgen's 1724 map and on 25 May 1789, the mill formed part of a sale, along with Hedgecourt Mill and Copthorne windmill according to the *Sussex Weekly Advertiser*. Later in 1832, John Saunders was the miller and he stayed until 1851, after which Richard Stanbridge took over. *Kelly's* 1859 directory lists a J. Stanbridge (also at Fen Place Mill) but this is the last reference.

A steep and narrow footpath from Furnace House leads down to the west side of the lake and below the embankment a tail race is just visible along with some old pieces of brickwork.

The disused Ifield Mill in 1974 prior to its restoration (TH)

The powerful waterwheel at Ifield Mill in the 1920's (VU)

The isolated Bewbush Mill and millpond (TH)

Ifield Mill in the late 1890's (TH)

Fen Place Mill in 1951

Hazelwick Mill in the early 1900's (PA)

Rowfant Mill in 1904 (NC)

Rowfant Mill as a house conversion in 1994

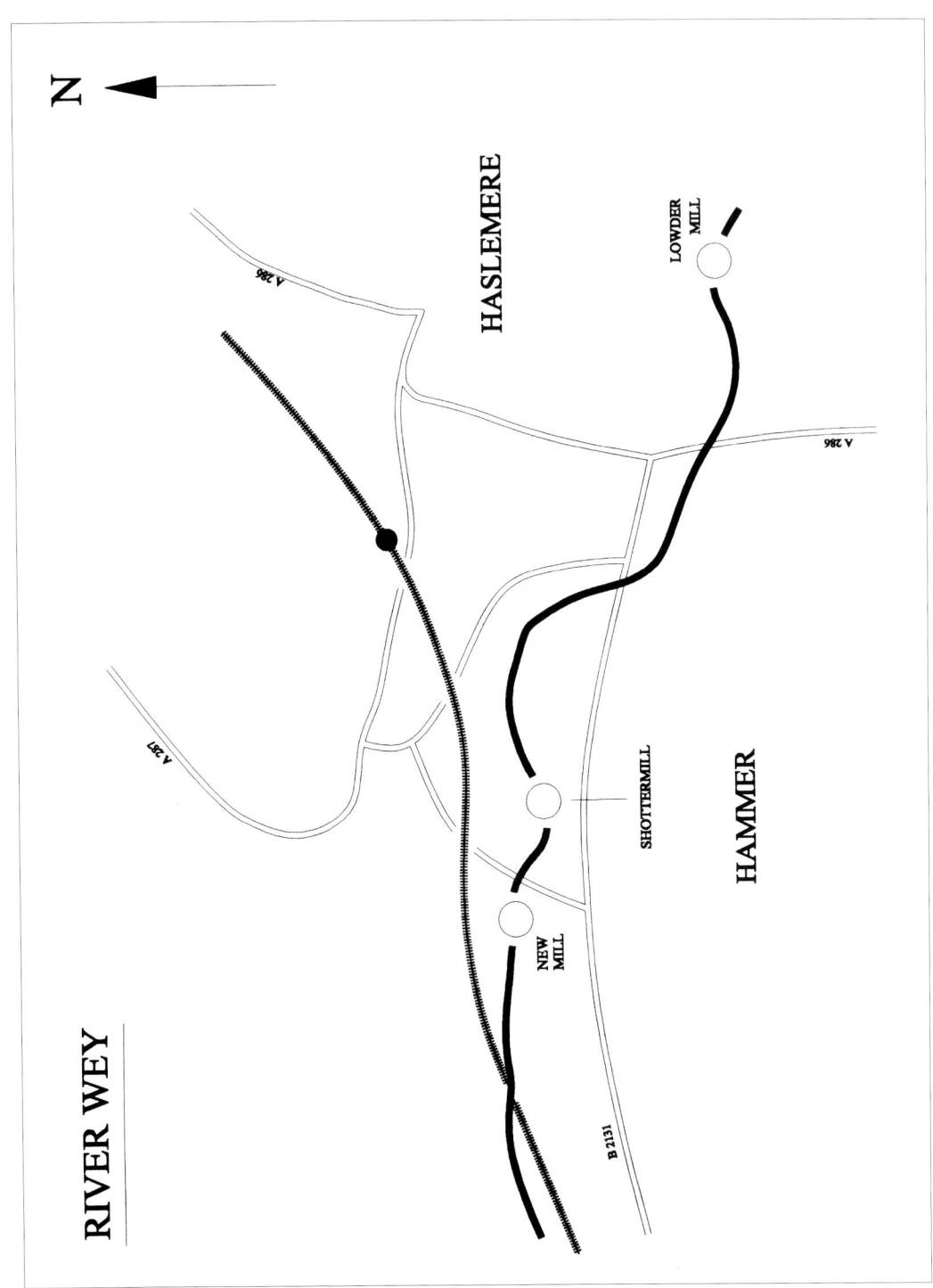

RIVER WEY

LOWDER MILL *North Ambersham*
Tributary to River Wey – SU 900 316 – Adjacent to Lowder Mill Road

This mill is first shown on John Budgen's 1724 map with no evidence of an older building situated on a small tributary of the River Wey, water supply must have been a problem and, to combat this, two pen ponds, plus a millpond, were constructed using ingenious earthwork excavation. Each of the ponds was built on the side of a hill, with a simple embankment supporting the lower side.

The 1851 *Water Resources Survey* gives the proprietor as J. Lucas, with the mill described as 'Flour and Grist, and containing 2-pairs of stones. The mill worked on average ten hours a day, which is not surprising considering the amount of stored water here, and produced 40 sacks of flour per week.

Little is known about the mill from 1851, and it is also not known when grinding ceased, although no miller is mentioned in the 1881 census. By 1939, the waterwheel and the associated machinery had been removed but the mill building survived. It is constructed in stone with some brick infilling, under a tiled roof with three storeys of no distinctive style, and is used as a garden store. The mill is partly hidden behind a small, but attractive mill house, and technically the whole site lies in West Sussex, with the county boundary with Surrey mered to the by-pass stream to the north of the mill.

NEW MILL *Haslemere*
Tributary to River Wey – SU 885 324 – South of Crutchmere Lane

This was the most westerly of the five Haslemere watermills and although the site has not been redeveloped, it is difficult to identify the position of the mill and its once expansive millpond.

A variety of trades were carried out here, ie. paper, flour and a skin mill. In 1801 it was known as Hall's Mill, after the resident paper maker, and it remained in the same trade for some time with Pewtress, Low and Pewtress in 1839 and James Simmonds in 1846. The 1851 *Water Resources Survey* lists that the mill contained 'Two Paper Mill Engines'. A directory entry in 1878 records Edward Dunce as 'a miller, grocer and postmaster' in occupation, while the 1873 Ordnance Survey 25" denotes the mill as 'Flour'.

By 1882, Edward Masters, a tanner, deer, buff and chamois leather dresser was using it in conjunction with the nearby Pitfold Mill, until trading ceased before the turn of the century, with the mill described as derelict in 1902.

The mill building was geographically in West Sussex, the county boundary with Surrey being adjacent to the northern bank of the old millpond. The site was later used as a pig farm, while in 1930, the millpond was filled in, although the mill building was still standing in 1964. Some of the footings can be found in thick undergrowth, while traces of the pond embankment can just be identified.

SHOTTERMILL *Haslemere*
Tributary to River Wey – SU 882 324 - On south side of Shottermill Rd

Until its closure in 1939, this was one of the last working mills in the area, but it has now been converted into flats. The old mill and its two feeder ponds are just in West Sussex.

It was a flour mill in 1826 in the occupation of William Oliver, in whose family it remained until it ceased working. Also, the windmill at Grayswood, was used by the Olivers until demolition took place in 1886. The 1851 *Water Resources Survey* records that the water mill only contained 2-pairs of stones. Major improvements took place in 1880 when the overshot waterwheel, 14ft in diameter by 7ft 6in wide, was replaced by a 'Little Giant' water turbine at a cost of £50. William Tomsett, who operated from the Ockford Works at Godalming, carried out the installation.

At the sale of the mill in 1940, the two millponds were purchased by the Haslemere Preservation Society, who later donated them to the National Trust. The mill and adjacent mill house are built in typical Victorian fashion, with yellow brickwork under a tiled roof, but have been converted into flats (the machinery was removed in the 1950's). A weatherboarded lean-to at the southern end of the building marks the former position of the overshot waterwheel, and the replacement turbine.

Lowder Mill as a garden store in 1939 (AS)

New Mill in 1850

The disused covered wheelpit at New Mill in 1964 (AS)

Shottermill closed in 1939 (AS)

Stone floor at Shottermill in 1939 (AS)

CONCLUSIONS

Reflecting back on all the watermills sites we visited in Sussex, it is disappointing that few are capable of grinding corn, or even having a waterwheel that can be turned by water power. The working mill at the Open Air Museum at Singleton, as good as it is, is a reconstruction of Lurgashall Mill. Indeed it does grind corn and look and smell right, even if the water is pumped electrically to the millpond, but its surroundings are such that you could believe you have walked into the past.

Woods Mill, at Small Dole, at the time of writing, is not in working order and has reconstructed machinery not in its original position. The building and the surrounding area are well worth a visit and it is hoped that the waterwheel is repaired back to working order soon. In East Sussex there is Bartley Mill, near Frant, which has, apart from the waterwheel, new machinery, but the mill is at present out of use. Park Mill at Batemans, is run by the National Trust and here there is better news. It still produces flour using machinery that is original and virtually intact. Michelham Priory Mill, also in East Sussex, is also of interest as it occasionally grinds corn, though the waterwheel is new. This in turn leads us to consider the mills in private ownership some of which could be capable of working with minor repairs. In East Sussex there is Hellingly Mill, with its new waterwheel and new pit machinery, renovated by an enthusiastic owner. There are sufficient remains at Sheffield Mill, with its ancient wooden layshaft and new waterwheel. West Sussex has five mills of potential restoration to grinding ability. Horsted Keynes Mill has a workable waterwheel, with the majority of the machinery, by watermill standards, quite modern. The mill has been restored, but off the beaten track and is on private land. Ifield Mill, near Crawley, is a testament to a group of dedicated and enthusiastic volunteers who have, over the last twenty five years, restored this once derelict and empty mill to a working condition powered by its original and exceptional wide waterwheel. The mill has been rebuilt from the first floor upwards, with the machinery either new, under construction or rescued from demolished mills. Warnham Mill, at Horsham, with its huge millpond, still retains its main machinery and workable waterwheel and needs preserving at all costs. Cobb's Mill, at Sayers Common, has an interesting range of buildings and unusual machinery with a waterwheel devoid of any buckets. How fascinating it would be to see the mill working again together with its auxiliary gas engine. Dean's Mill, at Lindfield, is perhaps the easiest mill to work again in Sussex, being the last mill run on a commercial basis. The mill is complete in every detail and with attention to the waterwheel, could easily work again. It would be tragic if this mill, possibly the best that Sussex has to offer, was allowed to deteriorate or be converted for any other use.

We can only appeal, through this book, to any mill owner to preserve what is left, bearing in mind how little remains of the two hundred and twenty watermills that formerly stood within Sussex.

GLOSSARY OF MILLING TERMS

Ark	A wooden bin for holding flour
Axle	A shaft linking the waterwheel with pit wheel
Bedstone	The lower and fixed millstone
Bin	Container for storing grain on the top floor of the mill
Bolter	Used for dressing or grading flour
Breastshot Waterwheel	Wheel where the water is projected against its centre.
Bridge Tree	Beam part of the hursting that supports the end of the stone spindle
Buckets	Fittings around the waterwheel that hold water
Burr Stone	Millstone quarried near Paris
Clasp Arm	Wheel whose spokes form a square around the shaft
Compass Arm	Wheel whose spokes radiate from the shaft
Cogs	Removable wooden teeth of a gear wheel
Corn	Grain or seed of any cereal crop
Corn Laws	Regulatory statute to control the import of foreign wheat
Crown Wheel	Cog wheel at the top of the upright shaft; drives ancillary machinery
Damsel	The device that vibrates the shoe and ensures and even flow of meal into the stones
Dressing	Process for sharpening worn millstones
Eye	Opening in the middle of a millstone through which grain enters
Floats	The wooden or metal paddles on a waterwheel
Grain Cleaner	Removes foreign material such as stones, earth etc from the grain
Grist	Animal feed ground at the mill
Headrace	Section of stream/river above a mill, frequently referred to as a leat
Hopper	Open topped wooden box through which grain passes to the millstones below
Hurst	Wooden framework supporting millstones and pit gearing
Jack Ring	Mechanism for raising or disengaging stone nut
Launder	Headrace in the form of a open trough carrying water to the waterwheel
Leat	See Headrace
Lineshaft	A horizontal shaft which transmits power to part of the mill by gearing or belts
Lucomb	A wooden cabin projecting from the bin floor containing hoisting gear
Meal	The ground product as it leaves the mill
Mill Bill	Steel implement used for stone dressing
Millstones	A pair of stones usually made of burr, peak and composition
Overshot Waterwheel	Water is projected over and past the top of the wheel
Peak Stone	Millstone quarried in Derbyshire
Pentrough	Trough conveying water to the waterwheel
Pinion	Small type of gear wheel
Pit Wheel	The first gear wheel inside the mill, affixed to the axle shaft

Pond Bay	The dam or embankment of a millpond
Race	Water channel above or below a watermill
Roller Mill	A machine using metal or porcelain rollers to grind flour
Runner Stone	The upper millstone revolving over the bedstone
Sack Hoist	Used to raise flour or grain up and down through the mill
Shoe	Tapering wooden trough leading from the hopper to the eye of the millstone
Smutter	Machine for removing diseased particles wheat
Spur Wheel	Affixed to upright shaft, above the wallower, engaging the stone nuts
Stone Nut	Pinion engaging the spur wheel from which the runner stone is turned
Stone	See Millstones
Stone Spindle	Shaft attached to runner stone
Thrift	Wooden handle of mill bill
Undershot Waterwheel	Water is projected under the wheel
Upright Shaft	Vertical wooden or iron shaft
Wallower	Small toothed gear wheel affixed to the upright shaft, driven by pit wheel
Wire Machine	Used for grading flour by a cylindrical wire sieve containing wire brushes

The Society For The Protection Of Ancient Buildings

A hundred years ago, windmills and watermills were sights so common that they were rarely remarked upon. Today, few are left, and their numbers are decreasing. The Society for the Protection of Ancient Buildings (SPAB) was founded by William Morris in 1877 to oppose the unsympathetic and harmful restoration of old buildings. In the 1920's there were moves to include windmills within the society and this came to fruition in 1931. Later, in 1946, watermills were included and a Wind and Watermill Secion of the SPAB was set up. Every year the Section receives notification of over 100 planning applications concerning mills that are listed by the DoE as being of architectural or historic interest. These sites are inspected and recommendations submitted.

Subsequently, many mills have been saved from damaging alterations and, in some cases, demolition. The Section has published guides to the philosophy of repair of mills, and for safe working in mills. Also, it organises a programme of events and two annual one day conferences and publishes a quarterly magazine and a series of occasional publications. The main responsibility of the Section is to encourage the sympathetic and proper repair of mills, wherever possible, to working order, so that authentic examples will survive for future generations to study and repair.

If this book has inspired the reader in an interest in the study of appreciation of watermills in general, they may wish to join the Wind and Watermill Section of the SPAB. For further membership details contact the Section Administrator at SPAB, 37 Spital Square, London E1 6DY.

BIBLIOGRAPHY

Bennett, R and Elton,	*History of Corn Milling (4 vols) Simpkin Marshall 1899*
Brunskill, R.W.,	*The Illustrated Handbook of Vernacular Architecture Faber & Faber 1972*
Gregory, F.W.,	*Sussex Watermills SB Publications 1997*
Haselfoot, A.,	*Batsford Guide to the IA of South East England Batsford 1978*
Reynolds, J.,	*Windmills and Watermills Hugh Evelyn 1970*
Reid, K.C.,	*Watermills of the London Countryside Charles Skilton Vols 1 & 2 1987/1988*
Stidder., D.,	*The Watermills of Surrey Barracuda Books 1990*
Stidder D. and Smith C,	*The Watermills of Sussex Barron Birch Vol 1 – East Sussex 1997*
Straker E.,	*Wealden Iron G. Bell &Sons 1931 (New Edition David & Charles 1969)*
Syson, L.,	*British Watermills Batsford 1965*
Vince, P.,	*Discovering Watermills Shire Publications 1978*
Vine, P.,	*Kent & East Sussex Waterways Middleton Press 1989*
Vine, P.,	*London's Lost Route to the Sea David & Charles 1986*
Vine, P.,	*London's Lost Route to Midhurst Alan Sutton 1995*
Vine, P.,	*London to Portsmouth Waterways Middleton Press 1994*
Watts, M.,	*Corn Milling Shire Publications 1983*
Wenham, P.,	*Watermills Robert Hale 1989*

INDEX

Adorian, Paul 24
Adversane 21
  Mill 21
Agate, John 105
Albourne 57
Aldingbourne Mill 14, 113, *125*
Allen, Alfred 27
  James 79, 82
Amberley 116
  Wild Brooks 28
Amies Mill 21
  Farm 21
Andrews, James 26
  Thomas 33
Ansell, John 80
Appledram Salt Mill 113
Ardingly 99
  Fulling Mill 17, 99
  Rate Book 99
  Reservoir 99
Arnold, John 105
  William 104
Arundel 12, 30
Ashby, Thomas 47
Ashington 47
  Mill 47, *59*
Atkings, Daniel 58
Ayling, Frank 82
  Jessee 82
  John 71
  William 71
Bailey, George & Son 100
Baker, Clara 34
  John 27
Balcombe Mill 99, *107, 108*, 142
Barnett's Mill 69, *89*
Barnham, Charles 116
Bartlett, George 31
  John 77
Baxter, John 122
Baybridge Canal 13
Bayntun, Charles 58
Bennett, Arthur 72
Bewbush Mill 14, 139, *144*
Bex Mill 69, 70, *84*
Bignor Mill 21, *35*
  Farm 21
Billingshurst 12, 23, 29
Bines Bridge 13
Birchenbridge Mill 22, *34*
Bishopstone Tide Mill 123
Birdham 16, 114
  Tide Mill 113, *124, 129*
Blaker, Arthur 76, 79
  &Son 76,80
Bluebell Railway 101

Bognor Lower Mill 70, *85*
  Regis 9
  Upper 70
Bolney 47
  Common 47
  Mill 47, 48, 52, *60*
  West Mill 48
Boniface, Richard 113
  William 122
Booker, George 100
  James 100
Boone, Thomas & Son 114
Bosham 16
  Mill 14, 114, *128*
Botting, Henry 23, 25
  John 32, 50
  William 25, 29, 57
Boxall, James 13
Brewhurst Mill 11, 22, 23, *35*
  Milling Co 23, 32
Bridger's Mill 52, 100, 102, *110*
Bristow, James 141
Broadbridge, Heath 23
  Mill 23, 30, 33
  Mill (Bosham) 115
Brown, Fred 75
  Joseph 28
  Thomas 49
Budd, John 79
Burgess, Frank 120
  Hill 9, 13, 51
Burton Mill 70, 71, *85*
Burtonshaw, J 56, 57
Bury 12, 30
  Mill Farm 30
Butcher, Joseph 21
Byerley, Thomas 118, 120, 123
Caffyn, Blaker 27
  Jacob 52
  Thomas 50
  & Sons 140
Cardiff 71
Carshalton 71
Carter, James 29
  William 29
Catt, Edmund 72, 83
  William 83
Chalcroft, William 78
Challen, Henry 70
  James 123
  & Sons 58
Chantry Mill 24, *36*
Chichester 9, 14, 16, 52, 72, 75, 79, 115, 118, 124
  Harbour 14, 16, 115, 116
Chiddingfold 12
Chitty, George 77, 79
Chorley Iron Foundry 121

Churchman, John 25
Clarke, Charles 27
Clarke & Hellyer 116
Clayton 54, 142
Cobb's Mill 10, *11*, 15, 48, 52, 53, 55, 151
Cobden, Richard 69
Cockaise Mill 101, *108*
Cocking 71
  Mill 71, *86*
Coltsford Mill 142
Coolham 56
Comber, John 101, 104
  William 101
Constable, John 31
Cook, Charles 58
Cooksbridge Mill 72, *87*
Cootes, John 26, 28
Copsale 48, 49
  Mill 48
Corn Laws 9
Coster's Mill 17, 70, 72, 80, *87, 88*
Coultershaw Mill 34, 73, 76, 80, *84*
Court Mill 49, 50, 61
Cowfold 22
Crawley 139-141, 143
Crowhurst, Henry 24
Cuckfield 49, 57, 58
  Upper Mill 17, 45, 51
  Parish rate Book 100
Cut Mill 115
Dadswell, David 140
Dale, Edward 74
  Thomas 33
Dann F. & Son 52
Dean's Mill 10, 13, 17, 100, 102, 103, *106, 109*, 151
Defence Schedules 24, 27, 32, 48, 54, 101, 103, 105, 123, 139, 140, 142, 143
Dell Quay 124
Devaynes & Harrison 78
Devils Dyke 56
Ditchling Common 13
Domesday Survey 15, 24, 31, 55, 69, 71, 73, 78, 81, 104, 105, 114, 121, 123
Downpark Brick Works 93
Duke, George 24
Drewitt, William 74
Drinkwater, Robert 119
  Ruth 119
  Woodruff 119, 122
Dunce, Edward 147
Duncton 73, 74
  Dye Mill 17, 75
  Mill 73-75, 77, *85, 90*
  Paper Mill 17, 75
Dunnings Mill 14, 139
Durford Mill 75, *91*
Durrant, Thomas 141

Eames, Leonard	74, 77	Gower, Jeremy	99	Hurston Mill	26
Easebourne	80	Graffham Common	69	Hurstpierpoint	48, 53, 55
East Grinstead	139	Grammes, James	54	Ifield Mill	10, 14, 15, 22, 51, 99, 139, 141, *143, 144*, 151
East Lavant	118	Thomas	54		
East Mascalls Mill	103	Gratwicke, John	55	Iping Mill	17, 78, 79, *93*
East Newell Mill	115, 119	Guyatt, James	123	Ireland, Maurice	73
Ebernoe	30, 34	Gwillim, James	34, 73, 76	Isle of Wight	120
Ede, Anthony	48	John	80	Jackson, Abraham	104
Edward	21	Hackett & Son	119, 123	Jeffrey, Henry	121
William	56	Halfway Bridge Mill	17, 76, 79	Walter	121, 123
Egremont, Lord	75	Hall, Rowland	75	Jenner, Robert	102
Eldridge, John	82	Hammond, Bros	29	Samuel	100
Thomas	82	Henry	74	William	100
Ellis, George	78	John	47	& Higgs	52, 100, 102
Emery, Edward	26	Hamlin, Frank	99	Johnston, Henry	77
Richard	26	John	51	Joyes, Edward	56, 76
Emsworth	116-118	Hammonds Mill	13, 14, 51, 55, *62*, 142	Henry	31
Town Mill	9, 11, 14, 116, 120, *133*	Hampton, Henry	47	Kennard, Thomas	100
Evered, Robert	82	Handcross	105	Keymer Mill	53
& Stratton	82	Hardham, Foundry	29	Killick, Amos	50
Everest, Peter	139	Lock	77	Henry	56
Ewell Lower Mill	82	Mill	16, 74, 76, 77, *91*	James	22
Farne, Charles	116	Harding, Richard	141	Mary	50
Fenner, Charles	116	Hardwicke, William	24	William	33
John	116	Harland, Anthony	101	Kiln & Clarke	122
Felbridge	143	Harris, James	24	Kingsford, Michael	122
Fen Place Mill	14, 139, 140, 143, *144*	Harting Mill	77	Kirdford Mill	9, 26, 27, *38, 39*
Fernhurst	72	Haslemere	14, 17, 147, 148	Knepp Mill	53
Fever, Henry	116	Haslingbourne Mill	83	Castle	53
Finskin Bros	32	Hazlewick	14, 140, *145*	Knight, Alfred	70
Fishbourne	117, 118	Haxted Mill	80	Knott, Parnell	124
Fittleworth Mill	34, 70, 75, 76, *90*	Hay, Benjamin	115	William	124
Ford, Thomas	28	Haywards Heath	9, 13, 100, 102	Lashmar, Peter	49
Fresh Mill	116, 120, *129, 130*	Heasman, Thomas	142	Thomas	50
Freshfield Halt	101	William	142	Lavington Common	69
Forge	103	Heath Mill	25, 27	Leake, John	141
Mill	103, *111*	Heaver, John	105	Leconfield Estate	79
Friend, Arnold	105	William	105	Legge, Henry	75
George	104	Hedgecourt Mill	143	Leigh Mill	53, *64*
Fuller, J.	56	Hellyer, James	116	Leland, Gen	30
Funtington	118	Hemsley, William	113	Leppard, Henry	47
Gadd, Edward	79	Henfield	54, 55	Lillywhite, William	77
Gainsford, John	51	Hermitage	9, 16	Lindfield	10, 13, 100, 102, 103, 151
Richard	51	Heyshott	70	Linfold Farm Mill	27, 33
Gale, Robert	141	Hide, John	40	Lintoft, H & E	21
Garstons Farm	47	Highbridge Mill	50-52, *63, 64*	Littlehampton Harbour	12
Garton, William	141	Hodson, Bros	122	Lodsbridge Mill	16, 76, 79, *94*
Gasston, Henry	104	Holden Bros	34	Lodsworth	69, 81
Gater & Son	120	Hole, Ernest	33, 53	London Health Centre	102
Gatehouse, Richard	115	Holland, Peter	57	Lowder Mill	147, *158*
Thomas	115	Holman, George	58	Lower Beeding	50
Gatewick	50, *62*	Horley Mill	140	Loxwood	22
Gibbon's Mill	16, 24, *36*	Hookers, Farm	53	Lucas, Charles	33
Glasgow	124	Mill	52, *63*	James	147
Godalming	17	Horsham	9, 10, 12, 21, 22, 32	Luckens, J.	104
Godley, John	51	Baptist Church	24	Lumley Mill	14, 117, *130*
Luke	101, 139	Town Mill	11, 25, 33, *37, 38*	Lurgashall Mill	10, 17, 72, 79, 82, *95*
Goldsmid, Abraham	141	Horsted Keynes	104, *108*, 151	Lusted, James	34
Goodcare, James	27	Mill	10, 13, 15, 104	Lywood, Bernard	102
Goodyer, Henry	119	Hunt, James	122	Manor Mill	54, *64, 65*
Gorringe & Son	52	John	77	Marchant, Henry	105
Gosden Lower Mill	50	Hurst Mill	52, 78, *93*	Markwick, Hugh	140
Upper Mill	50	Hurst Green	142	Marshall, William	21, 78

156

Masters, Edward	17, 147	
Matthews, William	71	
Meeres, Richard	82	
Merrett, John	51	
Merstham Mill	141	
Midhurst	13, 69, 72, 76, 81, 83	
North Mill	10, 17, 34, 72, 79, 80, *96*	
South Mill	17, 81	
Mid Lavant Mill	81, *96, 97*	
Miles, John	140	
Military Police	76	
Milland Mill	81, *96, 97*	
Mills, John	69, 80	
Moase, Henry	27	
Monk, James	69	
William	72	
Moores, Richard	69	
Morgan, Albert	56	
Morley, James	79, 142	
Muddle, Charles	47	
Newbridge	12	
Newick	17	
New, Charles	31	
New Close Farm	51	
New Close Flour Mills	51	
New Fishbourne Salt Mill	118	
New Mill	10, 17, 147, *149*	
New Slipper Mill	14, 16, 118	
Northbrook Mill	118, *131, 132*	
North Ambersham	147	
Nutbourne	27, 119	
Lower Mill	27, 28	
Upper Mill	27, 28, *39, 40*	
Tide Mill	119	
Old Fishbourne Mill	120	
Old Place Mill	25, *38*	
Old Salt Mill	16, 117, 120	
Old Shoreham Mill	54	
Old Slipper Mill	14, 116, 118, 120, 123, *133*	
Oliver, William	51, 147	
Osborne, Robert	27	
Oving Mill	14, 120	
Oxted Mill	142	
Pace, John	100	
Packham Family	47, 48, 52, 53, 55, 78	
Pagham Harbour	14, 122, 123	
Pallingham Quay	12	
Parham Lake	29	
Parsons, George	118	
Patching, Resta	25	
William	25	
Penfold, William	49	
Petersfield	13	
Petworth	13	
Pewtress Low & Pewtress	147	
Piltdown	102	
Pim, James	102	
Pitfold Mill	17, 147	
Port of London	9	
Porter, Richard	69	
Portslade by Sea	9	
Portsmouth	12, 115, 117, 122	
Potter, Benjamin	33	
Poynings	54, 56	
Prewitt, Richard	25	
William	23, 26	
Pulborough	9, 12, 28, 74	
Puttock, Thomas	79	
Rackham Mill	28, *41*	
Ratham Mill	14, 121, *127*	
Rayward, William	50	
Redmond, James	82	
Reeves Bros	70, 104	
Regent Iron Works	101	
Ripking, William	115	
River		
Adur	12, 13, 47-58	
Arun	12, 16, 21-34	
Medway/Mole	14, 139-143	
Ouse	13, 99-106	
Navigation	13, 103	
Rother	13, 69-83	
Navigation	73, 77	
Wey	14, 147, 148	
River Park Mill	81	
Rogate	75	
Roper, Walter	23	
Rose, Thomas	102	
Round Hill	49	
Rowfant Mill	14, 142, *145*	
Rowner Mill	15, 24, 29, *42*	
Farm	29	
Ruckford Mill	13, 15, 48, 51, 52, 55, *65*	
Rudgwick	32	
Runcton Mill	14, 122, *134*	
Ruthven, J.	69	
Saunders, Charles	121	
Sadler, Robert	124	
Scrace-Dickens Chas§	22	
Scull, Maurice	113	
Seaford College	75	
Seaton, William	143	
Sedgewick Manor	22	
Selbourne	13	
Selham	69, 72, 79	
Sharp's Mill	17	
Sharp, George	22, 26, 33, 70, 77	
Walter	77	
William	122	
Shean, William	117	
Shelley, Timothy	33	
Shepherd, Alfred	102	
James	117	
John	122	
Shermanbury	55, *66*	
Mill	55, 56	
Park	55, 56	
Sheriff's Farm Mill	104, *110*	
Shillinglee Mill	10, 17, 30, *43*	
Shoreham	54	
Harbour	54	
Shottermill	14, 148, *150*	
Shotter, Jabez	69	
Shulbrede Priory	73	
Shutt Mill	30	
Sidlesham Tide Mill	14, 16, 122, *126*	
Simmonds, James	147	
Slade, John	70	
Slaugham	105	
Manor	105	
Mill	105, *111*	
Slaughter Bridge	56, *67*	
Slinfold Mill	30	
Small Dole	57	
Smart, John	25	
Smith, Charles	113	
Douglas	58	
John	77, 119	
Samuel	28	
William	121	
Smith & Bell	78	
Southampton	117	
South Harting	83	
Spence, John	99	
Spring Mill	56, 57	
Stabler, Henry	56	
Stanford, John	24	
Stanbridge, John	140	
Stedham	82	
Mill	82, *92*	
Stephens, Edward	104	
Stevenson, William	140	
Steyne Food Co.	49	
Steyning	31, 49	
Stone Lower Mill	105, *111*	
Stone Upper Mill	106	
Stopham	13	
Bridge	13	
Storrington	31	
Mill	9, 31, *43, 44*	
Stoveld, William	27, 121	
Streeter, Richard	106	
Streatfield, Luke	139	
Stubbington, Edward	70	
Sullington	24	
Swanbourne Lake	31, 32	
Mill	31, *44*	
St. Leonards Forest	13	
Syon Abbey	116	
Terrell, Joseph	50	
Frank	27	
Terry, James	117	
Terwick	69	
Mill	82, *88, 89*	
Thomas & Co	120	
Thomsett, William	82, 119, 148	
Three Bridges	141	
Tide Mills	16	
Tidy, William	140	
Tilgate Mill	143	
Tithe Apportionment	30, 32, 34, 51, 55, 69, 75, 82, 83, 105, 115-118, 121	
Tollervey, Edward	117	
Trim, Alfred	24	
Trustlers Mill	57, *67*	
Tullet, Harry	25	
Turbines	16, 24, 25, 71, 105, 116, 121, 123	

Turner, Frank	74
Jacob	76
James	139
William	76
Twineham	47, 52, 68
Valebridge Mill	57, 58
Venables, Charles	78
Wadlow, Charles	31
Wakeford, Amos	115
Henry	81
Wanford Mill	15, 25, 32, *40*
Wantley Mill	32
Warner, William	80
Warnham Court Mill	10, 15, 17, 26, 32, *45*, 151
Warren Furnace Mill	14, 140, 143
Wassell Mill	30, 34, *40, 41*
Wateringbury Mill	31
Waterwheels	5, 25, 32, 51, 54-56, 58, 76, 80, 114
Breastshot	15, 17, 19, 27, 29, 82, 102
Flood	19, 23, 20
Undershot	19, 23, 26
Overshot	15, 17, 22-24, 27, 28, 31, 33, 47, 49, 53, 57, 70, 71, 74, 79, 81, 99, 100, 101, 113, 114, 119, 140, 142
Wedge, Earl	122
Wefare, John	32, 57
Wells, William	78
West Ashling Mill	17, 52, 118, 123, *134-136*
West, Charles	72
West Mill	57, 58, *61*
Westbourne	117, 118, 123, *137*
Mill	14, 123
Westhampnett Mill	117, 124, *137*
West Harting Mill	83
West Sussex Mill Co	122
West Lavington	17, 72, 80
Weyman & Co	81
Weald & Downland Museum	10, 11, 17, 72, 79
Wickham, Thomas	57
Windmills	
Cold Watham	30
Copthorne	140, 143
Cripplegate	28
Grayswood	148
High Salvington	47
Jill	54
Oving	121
Rock	47
Round Hill	49
Witt-Mann, John	78
Richard	80
Woodfield Fulling Mill	58
Wood's Mill	142
Worth	142
Wyattt, Francis	119
John	115, 119
Young, Henry	114
Thomas	114

SUBCRIBERS

1 Derek and Moira Stidder
2 Tom Stidder
3 Barry Stidder
4 Amy Stidder
5 Colin and Jane Smith
6 Stephanie Smith
7 Christianne Smith
8 Priscilla Nobbs
9 Vivien Uwins
10 Alan Stoyel
11 Peter Booth
12 David Morgan
13 A.F. Hill
14 D.T.N. Booth
15-16 K.R. Hemsley
17 S.W. Bartlett
18 Michael Yates
19 M Dafau
20 Christopher M. Sheffield
21 Tim Farrow
22 E.F. Goatcher
23 National Monuments Record Centre
24 Brian J. Lee
25 E.W. Henbery
26 J.G. Harman
27 Don F. Filmer
28 Mrs P. Bracher
29 Alan Barwick
30 Martin Bodman
31 A.J. Mitchell
32 Simon Janes
33 Mr & Mrs P.F.C. Parr
34 Helen Bryant
35 Peter J. Hill
36 D.A. Rogers
37 Mr A. Kite
38 T.M. Hine
39 Bob Simpson
40 Martin Watts
41. Nick Catford
42-43 Ann Turner
44 S. M. Biggs
45 Michael Oakley
46 Lawrence Stevens
47 Michael Harverson
48 John Day
49 A.R. Tulley
50 Maev Wilkinson
51 Nell Slocock
52 Sue Tombs
53 Mrs Bertha Terry
54 Douglas White
55-56 Mrs Carol Mackay
57 Adrian Thompson
58 George Vivian Hodges
59 Peter Wakefield
60 Allan Brigham
61 Michael Fuller
62 Niall Roberts
63 D.G. North
64 J. Kenneth Major
65 Derek Harvey-Piper
66 Alan Budd
67 P.S. Jarvis
68 Mr & Mrs D. H. Cox
69 Mr R.D. Ashton
70 R.G. Martin
71 Paul C Smith
72 J. Wooward-Nutt
73 Michael Edwards
74 Mr T Bishop
75 Martin Morris
76 Dr I. P. Crawford
77 Tony Bonson
78 David Bushell
79 Gerry Goodrich
80-94 West Sussex County Library Service
95 John W. Hill
96 J.M.H. Bevan
97 John Hayes
98 Elizabeth Blake
99 Mr L.A. Barber
100 Andrew Findon
101 Peter Dolman
102 P.R. Allen
103 Alan & Glenys Crocker
104 Alan F. Gifford
105 Jennifer Smith
106 Dr Cecil French
107 Edward Offen
108 Guy Blythman
109 Nigel Melican
110 Tim Ralph
111 S.D Robertson
112 P.M Wright
113 Rita J Ensing
114 Maurice W. Dunman
115 J Lee
116 Guildhall Library
117 Keith Hursey
118 A O Brown
119 Susan Gilbert
120 A R Bryan
121 R S Harlow
122 Alan R. Brown
123 Ian Mc Grath
124 Margaret Port
125-6 Bob Potts
127 Martin Newton
128 Dave Robbins
129 Geoffrey Mead
130 Will Steer
131 Tony Gowan
132 Mr Brian C.S. Eighteen
133 Geoff Holman
134 Mildred Cookson
135 Dr.Barry Job
136 R.Rowling
137 M.G Hardy
138 Geoffrey Starmer
139 Peter Stock
140 Muriel Coleman
141 N.Plastow
142 Miss R.E. White
143 Mr M Dillon
144 John Ambler
145 Kim Leslie
146 Miss E.M.S Hammond
147 Robin Wilson
148 Peter A Hazzard

149 Nigel Brown
150 Colin H Bartlett
151 Mr M Kemp
152 Mrs Clara B Stidder
153 Mr Alan R Stidder
154 Roy Bevan
155 C.T. Riley
156 John Tritton Pelling
157 Maurice Broomfield
158 Wessel Koster
159 Mrs M Simpson
160 Keith Preston
161 Andrew Borland
162 Margaret Baldock
163 David Butler
164 Beril Roberts
165 Derek Parsons
166 Geoff Morris
167 P. Buck
168 Richard Harris
169 David Tomlinson
170 David Nash
171 Nigel F. Divers
172 A.K. Walters
173 Lesley E. Brook
174 Roger Packham
175 Eric Groves
176 Roy Berry
177 S . M. Romain
178 Mr & Mrs R Proctor
179 Peter James
180 J.S.F. Blackwell
181 Jeff Sechiari
182 Brighton & Hove Borough Council
183 Hampshire County Library Service
184 R. A. Philcox
185 A.W. Pinchbeck
186 Helen Livingston & Frank Haskew
187 Clive Sayer
188 Tjerk Oosterhuis
189 John Bedington
190 Ton Meesters
191 Mr & Mrs G Mansfield
192 M Snow
193 Christopher Clark
194 Jonathan Minns
195 Peter Dixon
196 John Harvey
197 Mr M Daker

Birdham Mill

Stedham Mill in 1906 (VU)